PRAISE FOR
IAN SAMPLE'S *MASSIVE*

"[Peter] Higgs himself has proved almost as elusive as his eponymous particle. Until now. Ian Sample . . . persevered long enough to secure an interview with him, and the results are among the highlights of *Massive*, a lively account of the genesis of both the LHC and its most famous particulate quarry. . . . Sample has interviewed quite a few other leading scientists, too, and proves adept at prising insights from them. . . . We are kept hooked by [the book's] fine reportage, which makes clear the sheer achievement of the scientists and engineers who have built the LHC, the most complex machine ever made in the service of pure science. We learn, too, of the many theoretical concepts that will be probed by it."

—**Graham Farmelo**, *The Guardian* (London)

"[Sample] shows a keen eye for the personal equation, even while narrating large swatches of physics history."

—*Wall Street Journal*

"If you read just one popular-science book about the ubiquitous/ elusive particle this year, let it be this one. . . . According to our reviewer Andy Parker, Ian Sample's account 'could be the screenplay' for a Hollywood film about Higgs-hunting. Yet Sample is also careful with the science, giving credit to physicists other than Peter Higgs and avoiding the lazy assumption that particle physics begins and ends with the boson that bears his name."

—*Physics World*, **selected #3 on top 10 of 2010**

"Sample describes the competition and politics behind the experiments that have sought the eponymous boson. . . . He relates amusing anecdotes . . . [and] spins a good yarn. . . . To get a sense of the sociology and politics of high-energy physics, *Massive* is a good place to start."

—*Nature*

"An extraordinary book that tells the real human story behind one of the biggest science adventures of our time, managing to translate the complex concepts of particle physics into a real page-turner."

—Judges' announcement,
2011 Royal Society Winton Book Prize Shortlist

"This was my holiday page-turner: a clear and engrossing description of the physics of the Higgs boson (with surrounding weirdness), combined with a breathless account of the leap-frogging race for its discovery."

—**Dara O'Briain, comedian,** *New Scientist*,
Best Book of 2010: A Comedian's Choice

"A whirlwind tour of the discoveries that first revealed the subatomic world. . . . Like any good book, the excitement in *Massive* builds, culminating with the frenzied Higgs hunt at the end of LEP's run and more recently at the Tevatron at Fermilab in the US—both racing against time to bag the revered particle."

—**CultureLab, NewScientist.com**

"The definition of the Higgs boson and how it gives everything mass, and why it's important, comes alive for readers with little prior science background. Recommended for general-interest and science collections alike!"

—*The Midwest Book Review*

"[A] roller-coaster of a tale. Sample keeps the physics accessible, but the real pleasure is in the personalities and drama he reveals behind the hunt for one of the most elusive objects in the universe."

—*Publishers Weekly*

"Lively popular account of late-20th-century physics, physicists and their machines. . . . Quality science journalism."

—*Kirkus Reviews*

"The grand narrative in Ian Sample's book sweeps from the earliest speculations on the nature of matter; through the Second World

War and the dawn of nuclear weapons; the paranoia of the Cold War (during which science was seen as a source of national security); rival efforts by the US and Europe to lead the world in times of peace; and the eventual emergence of worldwide scientific co-operation. . . . *Massive* carries the reader through the epic using individual episodes from the lives of some of the participants."

—*Physics World*

"*Massive* is a tale of search and of discovery, of the hunt for a particle of high mass and very short lifespan called the Higgs Boson. . . . Go. Read. Enjoy."

—*New York Journal of Books*

"A compelling work of popular science, full of mind-boggling ideas and a real sense of the excitement of scientific discovery."

—*The Guardian* (London)

"British science writer Ian Sample's newest book, *Massive: The Missing Particle That Sparked the Greatest Hunt in Science*, relates the scientific and human history behind the LHC. . . . The focus on Peter Higgs may chagrin [other] researchers, who are probably weary of having their names lost in the glare of Higgs'. But that is an important example of the human story that Sample weaves in parallel to the scientific one."

—**Dr. Fred Bortz**, *Dallas Morning News*

"Sample is terrific at describing the various labs and their huge machines with their miles of underground tunneling and sudden collapses. Maxwell unified electricity and magnetism; Einstein, space and time; and now we're into supersymmetry and the weak force involved in radioactive decay. This book shimmers with possibilities."

—*Providence Journal-Bulletin*

"Science journalist Sample does an excellent job of capturing the history of the subject and the vivid personalities of some of the most famous living physicists. . . . *Massive* is an excellent nontechnical introduction to the history of modern particle physics right up to the present . . . Highly recommended."

—**Choice**

"The only thing more elusive than the Higgs boson, the so-called 'God particle' that physicists built a $10 billion device to capture, is Peter Higgs himself. . . . *Massive* has achieved the journalistic equivalent of capturing the particle: The story pins down how a young Higgs, disenchanted with the use of atomic physics for weapons, came to propose a new type of particle that solved a snafu in a theory on symmetries. . . . *Massive* offers a larger window into the minds that dreamed of the Higgs and the culture that shaped their search, not a text to explain the basics of modern physics, and is accessible for the curious science layperson."

—*Science News: Magazine of the Society for Science & the Public*

"Ian Sample does a masterful job of telling the tale of the quest for the Higgs boson (aka the 'God particle') in his new book *Massive: The Missing Particle That Sparked the Greatest Hunt in Science*. You don't need to know a thing about physics (though the author clearly does) to enjoy it. Sample has a talent for explaining things that are often obscured by mathematics (a kind of crutch, I think, for many scientists) in straightforward English prose. This skill, combined with the fact that Sample is a great storyteller with a great story to tell, make *Massive* an excellent read. You may not have liked science in school, but trust me when I say you'll very much enjoy the history of science in the hands of Ian Sample."

—Marshall Poe, *New Books in History*

"The physics book generating the most bloggy buzz in the latter part of 2010 would have to be Ian Sample's *Massive: The Missing Particle That Sparked the Greatest Hunt in Science*, about the as-yet undetected particle known as the Higgs boson. Detecting the Hiigs [*sic*] is the most immediate goal of the Large Hadron Collider, so it's a topic that's in the air at the moment. . . . This is, basically, a concise history of particle physics in the accelerator era, with a focus on the theoretical mechanism that accounts for the mass of the various particles making up the Standard Model. . . . engagingly written, well-researched, and a good, fast read."

—Chad Orzel, Uncertain Principles

"Often when you read a book about material very close to home, minor (or major) inaccuracies irritate. Happy to say this wasn't the case here, I didn't see any, and there was plenty of context and background which I didn't know and enjoyed reading."

—Jon Butterworth, *The Guardian* (London)

"The Higgs is a particle predicted by theoretical physics . . . that's thought to be responsible for endowing everything in the universe with mass. For theoretical particle physics that's a relatively exciting idea, but as a story the 'hunt' for the Higgs is pretty odd, because, though Sample artfully avoids saying it in such blunt terms, nobody has ever actually come close to finding one in nature. . . . What merit *Massive* does have lies in its provocative synthesis of all the reasons why particle accelerators are a great idea."

—*Good Men Project Magazine*

"Just in case you're looking for the perfect gift for the science enthusiast in the family . . . *Massive* turns the dry-sounding hunt for the Higgs boson into the equivalent of a scientific detective story that you can't put down. . . . Vivid detail and backroom chatter [make] *Massive* such a compelling read: it's about science as that science is being done, and we don't yet have all the answers—the Higgs continues to elude us. But for anyone curious about the story of the Higgs so far, you're not likely to find a better book than Sample's on the subject."

—Jennifer Ouellette, author of *The Calculus Diaries: How Math Can Help You Lose Weight, Win in Vegas, and Survive a Zombie Apocalypse*

"Ian Sample's *Massive* is a marvelous book and well worth reading by both researchers and the layman. In it, Sample describes the history and the personalities behind the search for the Higgs boson. He dispels the common simplifying myth that a single lone genius named Peter Higgs was the sole theoretical mind behind the idea. Instead, Sample gives appropriate credit to the many theorists who made equally critical intellectual contributions."

—Don Lincoln, Fermilab / *CERN Courier*

MASSIVE

The Missing Particle That Sparked the
Greatest Hunt in Science

IAN SAMPLE

BASIC BOOKS
A Member of the Perseus Books Group
New York

Hardcover edition first published in 2010 by Basic Books,
A Member of the Perseus Books Group
Paperback edition first published in 2012 by Basic Books

Books published by Basic Books are available at special discounts for bulk
purchases in the United States by corporations, institutions, and other organizations.
For more information, please contact the Special Markets Department at the
Perseus Books Group, 2300 Chestnut Street, Suite 200, Philadelphia, PA 19103,
or call (800) 810-4145, ext. 5000, or e-mail special.markets@perseusbooks.com.

Designed by Brent Wilcox

Library of Congress Cataloging-in-Publication Data
Sample, Ian.
    Massive : the missing particle that sparked the greatest hunt in science /
by Ian Sample.
        p. cm.
    Includes bibliographical references and index.
    ISBN 978-0-465-01947-2 (alk. paper)
    1. Higgs bosons.   2. Large Hadron Collider (France and Switzerland)   I. Title.
    QC793.5.B62S26 2010
    539.7'21—dc22
                                                                        2010023132

ISBN 978-0-465-05873-0 (revised paperback)
ISBN 978-0-465-02852-8 (first paperback)
ISBN 978-0-465-03169-6 (e-book)

10 9 8 7 6 5 4 3 2 1

For my parents

CONTENTS

PROLOGUE

The mountainside village of Crozet in eastern France has commanding views over miles of countryside. Villages and farmhouses dot the fields below, a few narrow roads meandering between them. Apart from a handful of modern buildings that sit in a huge ring on the landscape, there is nothing here that looks unusual.

But this is a far from ordinary place. Some of these surface buildings conceal deep shafts that reach down to the largest, most sophisticated machine mankind has ever built. If a giant tore it from the ground and stood it up like a hoop, it would reach more than five miles into the sky. To switch it on is to invite an electricity bill equal to that of a fair-sized city.

This is the home of the Large Hadron Collider (LHC), a multi-billion-dollar atom smasher run by CERN, the European nuclear research organization, on the outskirts of Geneva. More than twenty countries clubbed together to pay for this leviathan and took over a decade to construct it. Ten thousand scientists here and in laboratories around the world are connected to the information it churns out via a distributed computing grid that has been touted as a new model for scientific collaboration.

Inside the machine itself, fragments of atoms are whipped up to within a whisker of the speed of light and slammed together in head-on collisions. These orchestrated acts of violence are said to re-create conditions that prevailed in the first moments of the Big Bang, the cosmic eruption that gave birth to the universe. Amid these fleeting specks of primordial fire, scientists look for answers to the most profound mysteries of nature.

One of these mysteries, perhaps the most intriguing of all, has hung over scientists for nearly half a century. The frank admission is this: scientists cannot explain why stuff weighs what it does. They can get close—very close, in fact—but there is always something missing. And they know the reason why. Smash something to pieces, into dust, then atoms, then fragments of atoms, and you will eventually reach the smallest building blocks of matter. The baffling, perplexing truth is that scientists do not know why these particles—from which all else is made—weigh anything at all.

In 1964, a physicist working with pen and paper in his Edinburgh office stumbled on what most scientists believe is the answer to the mystery. Peter Higgs conceived of an invisible field that reaches into every corner of the cosmos. At the beginning of time the field lay dormant, but, as the newborn universe expanded and cooled, it came to life and made its presence known. In that moment the building blocks of matter flipped from weightless to weighty. The massless became massive. The consequences are all around us. They are the bedrock of our existence.

Without the field, our universe would be a frantic storm of particles hurtling around at the speed of light. The atoms and molecules we know would not exist. Cosmic material would never have clumped together to form galaxies, stars, and planets. There would be no familiar structure to the universe—nowhere for life to gain a first, tentative foothold.

A scientist at CERN once told me the field was like the snow that had fallen that night and settled on this idyllic French-Swiss landscape. Imagine a snowfield that goes on forever in all directions. Beams of light move through it as though they have skis on: they zip through the field as if it weren't there. Some particles have snowshoes and make less swift progress. Others go barefoot and are destined to trudge around at a snail's pace. A particle's mass is simply a measure of how much it gets bogged down in the field.

The Large Hadron Collider was designed to reveal once and for all the true nature of the field that Peter Higgs envisaged. The machine should create ripples in the field that appear as particles called "Higgs bosons." They are the snowflakes that make up our cosmic snowfield and the final proof scientists need to fully explain why stuff weighs anything.

CERN is not the only place hunting for the particle. On the outskirts of Chicago, scientists at Fermilab, home to the second most powerful collider in the world, also made the particle their top priority. For the two laboratories on either side of the Atlantic, the decades-long hunt has become the greatest race in modern physics.

There is more to finding the Higgs particle than pride. It is the only missing piece of the Standard Model, a set of laws that describe all of the known particles in the universe. But that is only the beginning. A growing band of scientists believe the Higgs particle will not only solve the mystery of mass, but open a portal to a hidden world of particles and forces we can only begin to imagine.

The elusive nature and profound importance of the Higgs particle led one Nobel Prize–winning physicist to give it a grandiose nickname: the God particle. As you will find, if you read on, few things unite physicists more than their disdain for the name. Their contempt is equaled only by the joy of newspaper headline writers, for whom it has become a savior of a very different kind.

This book is the story of how the universe got its mass, and how an idea written down in a notebook nearly half a century ago became the focus of a global, multibillion-dollar hunt involving thousands of scientists and the largest, most complex machines ever built. Whichever way you look at it, this story is massive.

1

Long Road to Princeton

The drive up to Princeton could take the better part of a day, and that was if you were lucky. The route followed the coastline up the eastern seaboard, looped around the broad expanse of the Chesapeake Bay, and went on to Washington, Baltimore, and Philadelphia before finally arriving in the town that was once home to the greatest physicist of all, Albert Einstein.

Peter Higgs packed some clothes and a folder full of research notes and went out to the car with his wife, Jody, and their six-month-old son, Christopher. He swung the suitcase in the back and had a long look at the road map. Satisfied with the directions, he pulled away, working north and east through the tree-lined streets and out toward the highway as the town eased itself to life beneath the spring morning sun.

It was March 14, 1966. Higgs, a physicist at the University of Edinburgh, had moved to Chapel Hill the previous year to spend his sabbatical at the University of North Carolina.[1] His work there had caught the eye of a prominent scientist, who invited him to give a seminar at Princeton's Institute for Advanced Study, one of the world's leading intellectual centers and the place where Einstein himself had spent much of his working life. The seminar was destined to be controversial: Higgs had proposed an idea that, if correct, could explain the origin of mass.

The trip turned out to be more than just another academic visit. It marked the beginning of a run of events that catapulted Higgs into the scientific limelight and set the stage for the greatest hunt in modern physics. Using multibillion-dollar machines occupying miles of underground tunnels, thousands of scientists have spent decades looking for the particle that formed the linchpin of Higgs's theory. Their mantra was simple: find the Higgs particle and the mystery of the origin of mass was solved.

For centuries, scientists had no idea that mass even had an origin, at least not in the modern sense of the phrase. The word "mass" described how much matter an object had, and matter was no more than a grand term for "stuff." A lump of rock had more mass than a loaf of bread (unless the baker was having an off-day), and that was that. The meaning of mass was so intuitive and tangible that no one seriously thought to question it.

Vague and incomplete notions of mass emerged in antiquity and were developed through the Middle Ages. Giles of Rome, a prominent theologian and one of the most influential thinkers of the late thirteenth century, took an important conceptual step when he distinguished between the dimensions of an object and the amount of matter it contained.[2] A block of ice, for example, clearly changed shape when it melted into water, evaporated into steam, condensed, and became frozen solid again. Yet the amount of matter remained the same, he said, whichever form it was in. The observation, which surely made for lively theological discussions about the Trinity, mirrors the modern definitions of volume and mass.

In the early fourteenth century, the Parisian philosopher Jean Buridan drew on the concept of mass when he described how throwing an object gave it an impetus that depended on how much matter it contained and the speed at which it was lobbed.[3] The sixteenth-century German astronomer Johannes Kepler took things further, arguing that planets stayed true to their orbits and didn't hurtle around space like scattered snooker balls thanks to the inertia arising from their enormous masses.

Despite the valuable work of early philosophers and astronomers, the term "mass" was not used systematically until 1687, when Isaac Newton laid the foundations of classical mechanics in a great but wholly impenetrable work, the *Principia*.[4] Newton said mass was a quantity of matter that arose from an object's volume and density. An object's mass governed its inertia, or how much it resisted being pushed around, and also how strongly it felt the force of gravity. With these definitions in place, Newton derived the basic laws of motion.

Newton had an intuitive grasp of mass and matter that went far deeper than he let on in the *Principia*. Worldly objects, he believed, were made up of countless tiny particles that were created by God and could never be destroyed. The particles came in different forms and sizes that stuck together to make different materials. All man could hope to do was fashion new shapes and forms from these conglomerations of vanishingly small particles.

Nearly twenty years after the *Principia* was published, Newton allowed himself to speculate on the nature of matter in his next great work, the more accessible *Opticks*: "It seems probable to me, that God in the beginning formed matter in solid, massy, hard, impenetrable, moveable particles . . . so very hard, as never to wear or break in pieces," he wrote.[5]

Newton's musings on matter were not so far off the mark. Today, scientists think of matter as being built up from a handful of particles that are almost indestructible. It took scientists more than half a century to identify the most basic building blocks of matter, which come together to form the innards of atoms. Variations of these give us the chemical elements of the periodic table: atoms that form metals, crystals, liquids, and gases and that intermingle to make an almost endless list of molecules.

Scientists call the ultimate building blocks of matter "fundamental" or "elementary" particles, and, by definition, they cannot be broken up into smaller pieces. The first was discovered in 1897 by J. J. Thomson at the Cavendish laboratory at Cambridge University.[6] Thomson, like many other physicists of his time, was intrigued by the nature of glowing rays that appeared when a voltage was applied across glass tubes filled with low-pressure gases. The rays moved from the cathode, which is the negatively charged electrode, to the anode, which is positive. What the rays were made of was a mystery.

Thomson began a series of experiments to investigate these curious "cathode rays." In one, he used a 15-inch glass tube coated at one end with phosphorescent paint. Thomson modified the anode by creating a slit in it, so some of the rays coming from the cathode would pass through it, making a bright spot when they hit the phosphor. His masterstroke was to build into the glass vessel a second set of electrodes that the rays passed through. When Thomson linked the electrodes up to a battery, he found that the spot darted away from the negative plate and toward the positive one.

Further experiments showed that cathode rays were composed of streams of tiny, negatively charged particles. Thomson named them "electrons," a term introduced by the Irishman George Johnstone Stoney twenty years earlier, and suggested they were ubiquitous ingredients of all the atoms scientists knew. Emboldened by his discovery, Thomson proposed the "plum pudding" model of the atom, so called because it pictured atoms as positively charged balls of matter (the pudding) dotted with tiny negative electrons (the plums).

It turned out that Thomson's atomic pudding was not what Nature ordered.[7] The idea fell apart when the New Zealand–born chemist and physicist Ernest Rutherford, based on his work with radium, announced the startling news that atoms were mostly empty. Instead, he said in 1911, almost all of an atom's mass was bundled up in a central, positive nucleus. Later that decade, Rutherford probed the nucleus more deeply and found evidence for a new kind of particle within, the positively charged proton.

By the mid-1930s, physicists had what they believed to be the main building blocks of matter. The nucleus of an atom was made up of protons and (except in the case of hydrogen, the simplest atom there is[8]) another type of particle, the uncharged neutron, discovered in 1932 by the English physicist James Chadwick.[9] Surrounding the nucleus was one or more negatively charged electrons. This interpretation was on the right track, but it was incomplete. Scientists later discovered that protons and neutrons were not elementary particles of matter at all. Unlike the electron, protons and neutrons were made up of even smaller particles called quarks.

It took a long time for physicists to accept that quarks were real, not least because no one had ever seen one. The American physicists Murray Gell-Mann and Georg Zweig first put forward the idea in 1964, though they hit on the theory separately.[10] They realized that the behavior of protons and neutrons made sense if each contained a trio of quarks. The proposal was still contentious when Peter Higgs visited the Institute for Advanced Study in 1966. Quarks were only widely embraced as true elementary particles of matter some years later.

In the half-century or so that followed Thomson's work on the electron, physicists identified around two hundred different kinds of particles, most of which were pairs or triplets of other subatomic ingredients.[11] The proliferation of particles was getting confusing, but order came in the mid-1970s with what must rank as the crowning glory of particle physics. Known as the Standard Model, a name so prosaic it is almost an offense, it explains all known matter from just a handful of truly elementary particles.[12]

According to the Standard Model, there are twenty-four fundamental building blocks of matter. Among them are six kinds of quarks (called up, down, top, bottom, charm, and strange), which come in three varieties depending on a property known as their "color charge."[13] The color charge can be red, green, or blue, but the names have no meaning in the visual sense. Quarks with different colors attract one another. That accounts for eighteen of the different particles. The remaining six are called "leptons," a family that includes electrons and ghostly, nearly massless particles called neutrinos, which pass almost unhindered through anything in their path. In our universe, the stable matter we know of is based on quarks and electrons.

The other particles described by the Standard Model are not building blocks of matter but are there to do other jobs. Four of them are responsible for transmitting forces of nature and are known as bosons.[14] The electromagnetic force, which is carried by photons—simply particles of light—is what keeps you from falling through the floor. Inside atomic nuclei, quarks are stuck together by the "strong force," which is carried by particles aptly named "gluons." Only particles with color charge feel this force, which is 137 times stronger than the electromagnetic force. Other particles, called W and Z bosons, carry what is known as the "weak force," which goes to work when certain radioactive elements decay.[15] One more particle completes the Standard Model, a theoretical particle predicted by Peter Higgs's theory, known as the Higgs boson.

You could be forgiven for thinking that the Standard Model wraps up all there is to say about the origin of mass. If all stable matter we know of is made up of quarks and electrons, then surely these elementary particles embody the smallest units of mass possible. By that reckoning, they are the origin of mass. If that was so, you could work out how much mass any object had just by totting up the contributions from all the zillions of quarks and electrons inside. It turns out it's not that simple.

When sums go wrong from the beginning, it usually means you are missing something. Here's a case in point. A proton contains two up quarks and one down quark. If you add up the masses of the three, the total comes to just 1 percent of the mass of the proton. A full 99 percent of the proton's mass is unaccounted for. The same thing happens with the neutron, which contains one up quark and two down quarks. If Newton had had the last word on mass—that it was simply a measure of matter—then adding up the masses of the individual quarks should give the right answers. But Newton knew only part of the story. The missing mass came from somewhere else.

There is more to mass than meets the eye. How much more became clearer in 1905, when a twenty-six-year-old Albert Einstein, while holding down a day job at a patent office in Bern, Switzerland, published a paper entitled "Does the Inertia of a Body Depend on Its Energy Content?" To cut to the chase, the answer is yes. Einstein showed that mass and energy are interchangeable, that mass can be considered a measure of how much energy an object contains. For the scientific establishment, the idea was a bolt from the blue, but it is an unavoidable consequence of Einstein's special theory of relativity.[16] The equation Einstein derived was $m = E/c^2$, where an object's mass equals its energy divided by the speed of light squared. Rearranged, the equation becomes the all-too-familiar $E = mc^2$, where the giant value of the speed of light, close to 300,000 kilometers (186,000 miles) per second, makes it easy to see how even tiny masses embody vast amounts of energy.

Einstein's revelation goes some way to explaining why the proton's mass is greater than the sum of its parts. The three quarks inside a proton account for only 1 percent of its mass, but they are held together by extraordinarily strong forces. The bulk of a proton's mass comes from the energy locked up in the movement of the quarks inside and the forces that bind them together. It leads us to a remarkable truth: any object you care to mention, from your pet dog to your cellphone, owes most of its mass to the intense energy it takes to keep it in one piece.

The interplay between mass and energy that Einstein highlighted is demonstrated beautifully inside the giant particle accelerators that physicists use to study subatomic particles. Slam two particles together at sufficiently high speeds, and the debris from the collision is likely to contain heavier particles than you started with. The energy released when the particles collide almost instantly condenses into fresh matter.

Between them, Newton and Einstein laid the foundations of our understanding of the nature of mass, but in the 1960s it was clear that something was still missing. Scientists could not explain where the fundamental particles got their masses from. It was this mystery that Higgs's theory seemed to solve. It gave scientists their best hope yet of fully describing the mass of everything they knew.

Peter Higgs arrived in Chapel Hill to set up home on September 6, 1965, having left Jody, who was heavily pregnant at the time, with her parents in Urbana, Illinois. At the university, he set about writing his first major paper on the origin of mass. On September 24 he was working in the library of his new department when he was called to the phone. His first son, Christopher, had just been born.

Higgs finished the paper in November and sent a copy for publication and a few more to physicists he thought might be interested. Though it wasn't clear at the time, Higgs's theory pointed to a critical moment in the birth of the universe. In the immediate aftermath of the Big Bang, the cataclysmic explosion that flung the universe into existence, the elementary particles were entirely massless. Then, a fraction of a second after the Big Bang, something happened: an energy field that permeated the fledgling universe switched on.[17] Massless particles that had been zipping around at the speed of light were caught in the field and became massive. The more strongly they felt the effects of the field, the more massive they became.

Most scientists concur that time began about 13.7 billion years ago, possibly with the first bang there ever was, but perhaps as just one of many bangs that occur in a cyclic process.[18] The universe, at first a microscopic ball of intense energy, was too hot for the laws of nature as we know them to yet emerge. But in the blink of an eye (had there been one around to oblige) the cosmos grew to the size of a beach ball and cooled just enough—to around ten thousand trillion degrees Celsius—for the Higgs field to come to life. In that eye-blink, the first building blocks of matter were tamed—made heavy and slow, like flies in soup.

The Higgs field is crucial to the structure of the universe and its ability to support life as we know it. Without the field, the elementary particles, the building blocks of matter, would behave like light. The chemistry we are familiar with would not be possible.[19] Matter would not have clumped together into the atoms we see today. Stars and planets would not have formed. The universe would be a lifeless wasteland.

At the heart of Higgs's theory was a new particle associated with the mass-giving field. The Higgs boson, in a sense, is the part of the field that is left over once the field has given mass to particles.[20] The best hope scientists have of showing Higgs's theory to be right is to show that the particle exists.

Not long after Higgs sent his paper out to academics, a letter arrived at his office in Chapel Hill. It was from Freeman Dyson, the English-born mathematician who had worked at RAF bomber command in World War II. Dyson had crossed the Atlantic at the age of twenty-three, clutching a letter declaring him the best mathematician in England. He had since become an eminent professor at Princeton's Institute for Advanced Study.

Dyson's letter was amiable and could not have been more flattering. He explained how he'd enjoyed Higgs's latest paper and that it made clear things he'd been puzzling about for some time. He asked if Higgs would give a seminar on his theory at the institute that spring. Higgs was astonished but accepted without a second thought.

Dyson's enthusiasm for Higgs's work didn't mean he was in for an easy time at the Institute for Advanced Study. The institute was home to some of the brightest physicists in the world, and some of them were certain to disagree with Higgs's theory. Renowned scientists had flocked to the institute since Louis Bamberger, an American philanthropist, had established it in the 1930s. Its most famous resident, Albert Einstein, who had died in 1955, had spent the last twenty-five years of his life there, trying to explain how the forces of nature were born. The Austrian-American logician Kurt Gödel was still there, redefining the limits of human knowledge. He and Einstein had been friends, though he had vexed Einstein by pointing out that his famous theories allowed time travel to be possible.[21] The father of modern computing, John von Neumann, was also at the institute, turning the mathematics of poker into a political strategy to win the Cold War.[22]

Robert Oppenheimer, the towering figure who had led the Manhattan Project to build the atomic bomb, had become head of the institute in 1946, only adding to the intimidating aura of the place. Oppenheimer was renowned for his short temper and sharp tongue and could be at his worst when he turned up for the weekly seminars that were held on the campus. It wasn't unknown for him to bully less self-assured speakers, quizzing them relentlessly and correcting them before they had a chance to respond. It was a character trait Dyson despised, and it occasionally triggered rows between the two men when seminars were over. "Oppenheimer always tried to tell you what you would have said if you were as clever as Oppenheimer," Dyson recalled.[23]

As Higgs drove on, his mind wandered to the talk he would give the following day. The audience would be unlike any he had spoken to before. When Higgs returned his attention to the road, he was gripped by a surge of panic.[24] Afraid he was about to lose control of the car, he pulled over. On the side of the road, he took a few deep breaths and tried to regain his composure. Higgs had just seen a road sign. The exit for Princeton was only a mile ahead. He was nearly there.

The Institute for Advanced Study sat amid 800 acres of landscaped gardens a mile outside of Princeton. Instead of driving directly there, Higgs took a detour into the town and parked his car. He found his way to a post office and had a word with the clerk. From behind the counter, the man produced a first-day cover of the violet eight-cent stamps that had been issued that day in commemoration of Einstein. Einstein, it turns out, had been born on that date in 1879, so the post office had carefully chosen both the date and the place of issue for the stamps. Each carried a picture of the great physicist taken twenty years earlier by Philippe Halsman, a family friend who had served time in an Austrian prison for killing his father on an Alpine hike. The stamps irked Higgs because they referred to Einstein as a "prominent American." Although Einstein had taken U.S. citizenship in 1940, Higgs considered him a European at heart. Nonetheless, he thought the gift would go down well with his friend and mentor Nicholas Kemmer back in Edinburgh and he duly posted it back to Scotland.

It was approaching evening when Higgs pulled up at the institute, where he and Jody met Freeman Dyson. The three got on well and Higgs soon forgot the nerves that had overcome him on the drive there. When they were done chatting, the Higgses went to their lodgings and collapsed into a much-needed sleep.

Higgs's talk was scheduled for 4:15 P.M. the following afternoon. When he arrived he saw that Dyson himself was down to speak first. The talk was highbrow, on the stability of matter, but it addressed a simple enough question: how is it that the objects around us stay intact, considering that they contain countless particles held in place by finely balanced yet extraordinarily powerful forces? Why doesn't this book, with the enormous amount of energy locked up in its atoms, suddenly tear itself apart? Why don't your clothes spontaneously explode into a zillion subatomic fragments?

Dyson wrapped up his brilliant lecture and opened the floor for questions. As expected, the audience was sharp and challenging. Their skills had been honed by something of a tradition at the institute, weekly lectures with the intriguing name of Shotgun Seminars.[25] At these lectures, no one knew who the speaker would be until a name was drawn from a hat. The presentations were popular, but there was a catch: every member of the audience had to put his name into the hat and be ready to give the seminar if his name was the one drawn. Everyone was fired up to speak, but those who were not chosen were equally fired up to grill whoever took the floor.

When the audience had exhausted its supply of questions, Dyson called a tea break and announced that their guest, Peter Higgs, was up next. As a special guest, Higgs knew he would be presenting. He joined the crowd for refreshments and over a cup of tea fell into conversation with a German physicist named Klaus Hepp. The two had met once before at a summer school in Scotland in 1960. As the two sat chatting, Hepp mentioned a paper that was about to be published by three highly regarded scientists.[26] It was, he was sorry to say, a body blow for Higgs's theory. "There's no doubt about it," Hepp said, as the two made their way back to the lecture hall. "You've got something wrong."

At least Oppenheimer wasn't around. Higgs had no idea, but the bullying director was gravely ill with cancer and was to formally step down three months later.[27] At the podium, Higgs collected his papers and, step by step, took the audience through his theory. Dyson listened intently. He thought Higgs's work was beautiful.[28] When their guest had finished speaking, several hands shot up around the room.

Although Higgs was anxious about the seminar, there was an underlying confidence in his manner. He knew the equations in his theory so well he could feel their deeper meaning. He was sure the ideas he put forward were sound. That didn't mean they were true, of course. Many things that are theoretically possible are not realized in nature. But if the theory was flawless, it was at least a contender for describing the origin of mass.

The questions from the audience were insightful, probing, and critical, but none exposed any mistakes in Higgs's logic. His theory had passed its most challenging test yet. Dyson thanked Higgs for speaking and closed the session, delighted that the talk had gone well. Later, Higgs heard that Arthur Wightman, a leading physicist in the audience, had told his colleagues they had better go back and check their "proof" that contradicted Higgs's theory. He had believed every word Higgs had said.

The next day, after dinner with the Dysons, Higgs took to the road again. A second invitation had arrived from Harvard University, where the prominent and playful physicist Sidney Coleman worked, and Higgs had agreed to drop in for an open discussion before heading back to Chapel Hill. The talk had been scheduled for the afternoon, which wasn't unexpected. Coleman famously missed morning appointments and once explained his failure to give a 9 A.M. lecture by protesting that he couldn't possibly stay up that late. Coleman was hoping to have some fun with Higgs.[29] He later confessed that he'd told his students some idiot was coming to see them. "And you're going to tear him to shreds!"

The mauling never happened. At Harvard, Higgs's talk turned into an enthusiastic discussion that everyone participated in. Once more, the theory stood up to scrutiny. If the audience began with hopes of pulling it apart, they finished with a sense of intrigue. Higgs's theory was a watershed moment in physics, one of those crucial steps that opens a door to a new world where discoveries are there for the taking.

Science has a long history of squandering brilliant ideas. They come at the wrong time, or they are clumsily explained, or they don't get noticed by the right people. Any one of these can relegate a profound leap in understanding to obscurity and deaden the progress of science. In a trip lasting less than a week, Higgs had made sure that his theory wouldn't vanish without trace. Gradually, other physicists started trying to make sense of it. They started to talk of the Higgs mechanism, Higgs fields, and the particle whose existence would prove it all to be real, the Higgs boson.

That autumn Higgs returned to Edinburgh and threw himself into his work with renewed vigor. An obvious question mark hung over his theory. Was it more than just a brilliant idea? Could it be a neat trick that nature had passed up on? He needed to take the idea and show how it worked in the real world. As it was, the theory was light on details. It showed how weightless particles might become massive in the early universe, but physicists had a whole zoo of particles to explain. Some of them had mass and others didn't. The theory didn't make clear which particles the Higgs field gave mass to and why.

The answer was destined to be one of the quests of the century. The handful of physicists who truly grasped Higgs's idea suspected that it paved the way to what many considered the greatest goal of their discipline. Toward the end of his life, Einstein became obsessed with proving that the different forces of nature, such as electromagnetism and gravity, were originally part of one all-encompassing superforce that existed only fleetingly at the birth of the universe. Since that time, physicists have wondered if there might be a "Grand Unified Theory," one that achieves Einstein's dream. Higgs's theory explained how nature could take all of the particles in the universe and, at a stroke, order some of them (the constituents of matter) to become heavy, while leaving others (like the photon) to remain massless. To physicists it seemed like a hint: that if they only dared to take Higgs's theory a little further, they might finally see how to reconcile all of nature's forces.

2

Shadow of the Bomb

The towering masts outside the Marconi Company's headquarters in Chelmsford, Essex, must have looked rather daunting to Dame Nellie Melba, the Australian prima donna whose silvery vocals had made her the star of Victorian opera houses. Nellie, who had performed everywhere from Manhattan to London and the Continent beyond, had only agreed to the visit reluctantly.[1] She was dismissive of the company's new "wireless" radio sets and couldn't fathom why hordes of people were clamoring for the "magic playboxes." Nevertheless, the diva was nobody's fool. The prospect of an event that promised £10,000 and global media coverage was not to be sniffed at.

If the singer was dubious to begin with, her host, Arthur Burrows, did little to assuage her fears. Gesturing to the sky, he explained how, from the tops of the giant radio masts hundreds of feet above them, the Dame's voice would carry for hundreds, maybe thousands, of miles to eager audiences in the European cultural capitals of Paris, Berlin, Madrid, and who knew where else.

The singer, evidently unfamiliar with the principles of the new-fangled technology, looked warily at the antennae and then to her guide. "Young man," she said, "if you think I'm climbing up there, you are greatly mistaken." Later that afternoon, on June 15, 1920, Dame Nellie became the star of the world's first concert to be broadcast over the airwaves.

The half-hour recital went ahead in a former packing shed that had been converted into a makeshift studio. There, clutching her handbag and leaning into the microphone (a telephone mouthpiece fitted with a horn made from an old wooden cigar box), the world's leading soprano began to sing to the accompaniment of a small grand piano. Word got around that the star was in town, and, outside, the local constabulary struggled to keep a heaving crowd of admirers at bay.

The broadcast was a tremendous success. At London's Imperial War Museum and also at the Crystal Palace, home to the Great Exhibition of 1851, engineers had set up wireless telephone sets where passers-by could come and listen. Dame Nellie sang Herman Bemberg's "Nymphes et Sylvains" and the "Addio" from La Bohème and finished with the United Kingdom's national anthem. Her voice carried further than Burrows dared hope, as far as Persia and Newfoundland. Later, she told reporters it was the most wonderful experience of her life. The stunt, bankrolled by Lord Northcliffe, the owner of the *Daily Mail* newspaper, fueled an upsurge of public interest in radio.

Radio was firmly established in Britain by the time Peter Ware Higgs was born in Newcastle on May 29, 1929. His parents, Thomas and Gertrude, had moved to the city from Bristol a year earlier, when Peter's father had taken a job with the British Broadcasting Corporation. Thomas worked on the technical aspects of radio broadcasts just like Nellie Melba's.

Broadcasting was a revolution born of revelation. It ranks as the most striking example in history of how understanding the laws of nature can provide a springboard for technological progress and transform global society. If ever there was an argument for the importance of basic blue skies research, the story of broadcasting is it.

It all began with Dafty. The nickname was given to the nineteenth-century Scottish physicist James Clerk Maxwell by schoolmates in Edinburgh who ridiculed his strong Dumfriesshire accent and "rustic and eccentric" appearance (his clumpy homemade shoes didn't help).[2] But Maxwell was a genius. His work paved the way for modern communications, from radios and digital television to cellphones and satellite navigation.

When we celebrate great figures for their most visible legacies, we risk losing sight of their real accomplishments. The technology Maxwell's work enabled can easily overshadow the conceptual breakthrough that underpinned his achievement. The Scotsman embraced two ideas that were so successful they have shaped physics ever since. Both are central to the Higgs story. The first was Maxwell's use of "fields," which at the time was a controversial and unsubstantiated concept.[3] The second was his way of working. Maxwell showed that the secret to great discoveries was finding connections between seemingly separate natural phenomena. The idea turned out to be so fruitful that it has become a working philosophy for scientists.

In 1860, at the age of twenty-nine, Maxwell was made redundant from his job as professor of natural philosophy at Marischal College in Aberdeen (despite having married the principal's daughter). His post was axed when the college merged with neighboring King's College in Aberdeen to form the city's university. After failing to secure a similar position at the University of Edinburgh, Maxwell made his way south, to King's College London, to take up the vacant chair of natural philosophy there.

Maxwell's spell at King's was arguably the most productive period of his life and it was there that his work with electricity and magnetism flourished. His interest began with experiments that Michael Faraday had done years earlier in a ramshackle lab at the English capital's Royal Institution. In one elegant demonstration, Faraday had taken a wire coil in one hand and a bar magnet in the other. When he moved the magnet inside the coil, an electric current sprang to life in the wire. When he held the magnet still, the current vanished. Faraday described the effect by saying that a moving magnet created an electric "field."

Faraday's experiments suggested electricity and magnetism were somehow connected, but no one knew how. Maxwell started writing down equations. His aim was to see if there was a way to link the two mathematically. The formulas he derived showed not only that electricity and magnetism were related but that they were two sides of the same coin.

Maxwell's equations look sparse and obscure on paper, but plug numbers into them and they take on a life of their own. They show that a moving magnet creates an electric field, but that's just the beginning. The newly created electric field produces a magnetic field of its own, which creates another electric field. And on it goes. The ripples in the electric and magnetic fields egg each other on, spreading out into space in perpetuity.

Maxwell wasn't finished. He was intrigued by these ripples of electromagnetic field and wondered how fast they spread out from their source. The answer was staggering: they traveled at the speed of light. On seeing that number drop out of the calculations in his notebook, Maxwell must have felt a rush of excitement known only to those who are first to uncover profound truths about nature. His work suggested that the ripples of the electromagnetic field and light waves were essentially the same thing.

Maxwell's work was instrumental in putting the concept of fields on a sure footing. As such, it laid a necessary foundation for Peter Higgs's work, which uses fields to explain the origin of mass. Later, Einstein gave particular credit to Maxwell for opening scientists' eyes to the importance of fields, commenting: "This change in the conception of reality is the most profound and the most fruitful that physics has experienced since the time of Newton."[4]

Scientists learned another great lesson from Maxwell.[5] In looking for a connection between two different things, electricity and magnetism, Maxwell uncovered a deeper truth about nature. What started off as an attempt to explain Faraday's observations ended with a theory of light. From there came the discovery of other kinds of electromagnetic waves, including the radio waves that carried Nellie Melba's voice over Chelmsford and brought the first fuzzy images to 1920s' television sets.

Maxwell's work posed an immediate challenge for scientists thinking about matter. At the time, Newton's view of reality prevailed; scientists broadly agreed that everything in nature could be explained in terms of matter in one form or other. Believe this and there was no need to resort to fields. From Newton's laws alone, you could describe all of the material and motion of the cosmos as though it were a giant mechanical device.

The obvious clash came in the different interpretations of light. Newton said a beam of light was a stream of tiny particles, or "corpuscles," but to Maxwell it was a wave. That led to the question of what was waving. What was the nature of the electromagnetic field? It was a question Maxwell himself was at a loss to answer. The reaction of scientists to the problem reveals how hard it can be to overturn well-rooted beliefs in science. Their response was to propose the existence of the "ether," a bizarre form of matter that pervaded the universe.[6] Light waves, they said, must be pressure waves in the ether, akin to sound waves in air.

To believe in the ether must have taken some doing. Scientists knew that sound waves moved faster in liquids than in air, and faster still in solids. They also knew about the incredible speed of light. If light was caused by pressure waves in the ether, then it followed that the ether must be a truly exotic substance. On its own, that wasn't so much of a problem. But at the same time, the ether was invisible and didn't obstruct the planets as they hurtled around the heavens. Even the slightest drag on these celestial bodies would cause them to slow down and ultimately spiral gracefully into the sun.

With hindsight, the ether seems like a desperate attempt to patch up a theory with holes in it, but we shouldn't feel too smug. Elaborate attempts of this sort are alive and well and papering over the cracks in our understanding of science today. At best, they serve a worthwhile purpose, ensuring that good ideas are not dismissed because they are incomplete. At worst, they are time-wasting dead ends that do to scientific progress what body shots do to zombies: they might not stop it in its tracks, but they make going forward painfully slow.

The wonders of nature are often held up as proof of God's great work, and so it was with the ether.[7] If it was real, the ether was inconceivable in size, entirely unblemished, and had properties that were hard to reconcile. For the religious-minded, and Maxwell was among them, only God could pull off that kind of job. Lord Kelvin, a prominent scientist at the time, was in no doubt that light was transmitted by pressure waves in the ether, though he was also convinced radio had no future and maintained that "passenger planes" were a harebrained idea that would never get off the ground.[8] Kelvin personified two of the strangest paradoxes in science: that at any moment in time, the latest theory of nature will almost certainly turn out to be wrong; and that scientists are often the last people to see how the discovery of today will become the technology of tomorrow.

Maxwell's discovery of the nature of light firmly established the concept of fields, thus laying a major foundation for Higgs's theory. But a far more dramatic turn of events was needed before Higgs could make his breakthrough. The quantum revolution began twenty years after Maxwell's death and ended its first phase around the time Higgs was born. In no other era had physics been so bewildering and controversial.

The greatest of scientific revolutions can start with the most unlikely experiments. In the case of quantum physics, the experiment in question involved someone staring at an oven as it heated up and watching it glow in different colors. For the most powerful and counterintuitive idea in twentieth-century science, it was a less than glamorous beginning.

In the hands of an ordinary scientist, the experiment could have gone nowhere. But Max Planck, a mustached physicist at Berlin University, was far from ordinary. He was obsessed with understanding the workings of nature. Einstein described Planck as being driven by a "hunger of the soul."[9] His dedication to science, it was said, was like the passion of a man in love.

In his autobiography, Planck elaborated, writing: "It is of paramount importance that the outside world is something independent of man, something absolute, and the quest for the laws which apply to this absolute appeared to me as the most sublime scientific pursuit in life."[10] Perhaps more than anyone, Planck understood that the reality we experience day to day is the culmination of countless laws that play out in the hidden realm of the microscopic.

Everything around us absorbs and radiates energy in the form of electromagnetic waves. When humans give off heat, the energy is released as infrared radiation. These waves are invisible to the human eye, but they can be picked up by thermal cameras, goggles, and the like. Police now use such equipment to track criminals in the dark, and firefighters use it to rescue people in smoke-filled buildings. But such technology was not around when Planck was breaking new ground in the field. The experiment that intrigued him used ovens specially designed to study thermal radiation.

Planck didn't have to do the experiment himself. He was a theoretical physicist and was able to grab the data from friends working down the corridor in the university's laboratory. His job was to find meaning in their results, and to do that he had to make sense of the rainbow of colors that emanated from the ovens.

As the ovens heated up, they glowed, at first a dull red, then bright red, then orange, yellow, and eventually a brilliant white. The hue depended on the temperature of the oven to such an extent that you could tell how hot it was from its glow alone. To Planck, the changing colors pointed to "something absolute": a fundamental truth about nature.

For the colors to change as they did, the ovens had to be giving off light with ever shorter wavelengths as they heated up. Red light has a longer wavelength than orange, which has a longer wavelength than yellow light. At the same time, the range of wavelengths emitted got broader. At the hottest temperatures, the ovens glowed a bluish-white because all the colors of the rainbow were being given off at once.

Now and then science takes a leap forward through nothing more than intuition, and Planck had intuition in spades. One afternoon, he came up with a mathematical formula that fitted the experimental results from the ovens almost perfectly. Planck's equation turned out to be so accurate that, as scientists improved their equipment, the data from the glowing ovens matched his predictions ever more closely.

Planck's next job was to work out the secret of his success. He knew that the amount of light an oven could give off was limited by its temperature. That's because the oven's temperature is a measure of how much energy it holds, and it loses some when it gives off light. The hotter an oven gets, the more energy it has available to radiate. Planck realized that, counter to accepted wisdom, shorter wavelengths of light must carry more energy than longer ones. When the oven was warm it could only radiate red light, but at higher temperatures it had enough energy to glow orange, yellow, or even bluish-white.

The true magic of Planck's discovery came next. He realized that his formula only worked if energy was radiated from the ovens as a stream of tiny pulses, or packets. The amount of energy in each packet depended on the color of the light. Packets of red light carried less energy than packets of blue light. Planck published his results in 1900, and, in time, the packets of energy became known as "quanta." It wasn't a bad year for a man who at the age of seventeen had been advised to steer clear of physics because all the real work had been done.

Planck was an accidental revolutionary. Neither he nor anyone else immediately grasped how profound a step he had taken. He saw his theory as more of a mathematical convenience than a true representation of the physical world. It is easy to see why he was cautious. Planck's theory flew in the face of Maxwell's supremely successful work, which had described light as a continuous wave of electromagnetic energy.

It took Albert Einstein, still in his early twenties, to understand the full significance of Planck's work. In 1905, Einstein realized that if he interpreted quanta literally he could explain one of the mysteries of the time, the so-called photoelectric effect.[11] Lab experiments showed that shining light onto a lump of metal could make an electric current flow from it and around a circuit. What was curious about the effect was that it worked well with violet light, but was hopeless with red light.

Einstein reasoned that electrons on the surface of a piece of metal stayed put unless they absorbed a certain amount of energy. They could only start moving if the light shining on them carried quanta of energy that were sufficiently large to overcome the metal's clutches. You could point floodlights at a lump of metal and fail to shift any electrons if the light was the wrong color, because the quanta carry too little energy. It is like trying to win at a fairground game by lobbing Ping-Pong balls at the target. Throw as many as you like, but the game is rigged: the target won't budge until you hit it with something more hefty.

Einstein was awarded the Nobel Prize for using the concept of quanta to explain the photoelectric effect. His work made clear that the idea of energy being divided into quanta must be taken seriously, and it quickly became the central thrust of theoretical physics. Four years after explaining the photoelectric effect, Einstein argued that quanta of light had momentum, making them particles in their own right. Scientists wrestled with the idea of quanta for three decades before it was turned into a meaningful theory that explained matter on the atomic scale.

Radical and counterintuitive as these ideas were, physicists knew they had to grasp the meaning of the quantum if they wanted to understand matter. The work of Planck and Einstein showed that the atomic world was governed by laws that were completely different from the ones Newton had discovered for the macroscopic world that dominated our daily experience. Newton's laws work fine for big things like cars and cannonballs, but strange and nonintuitive rules govern the realm of particles. The building blocks of matter simply cannot be understood without understanding the rules of the quantum world.

The structure of the atom as perceived today was still being fleshed out when quantum physics came on the scene. Proposals from Ernest Rutherford and the Danish physicist Niels Bohr suggested that atoms had a hard nucleus encircled by electrons in concentric orbits. In 1913, Bohr realized that a quantum interpretation of electron orbits allowed him to explain the colors of light absorbed and given off by hydrogen gas. It was a very specific piece of work, but it bolstered physicists' confidence that the quantum was key to understanding the structure of matter.

For more than a decade, physicists' work on quanta was fragmentary and piecemeal. What they needed was a full and general quantum theory that explained the behavior of any particle in any atom or molecule they cared to observe. This grand achievement emerged over an intense four-year period that began in 1925 and ended with the establishment of "quantum mechanics," the application of quantum theory to the goings-on of the atomic world.

Werner Heisenberg, a twenty-four-year-old physicist at the University of Göttingen, Germany, was first to make serious headway. Toward the end of May 1925, Heisenberg was suffering badly from hay fever and asked his supervisor, Max Born, for a fortnight's leave. He made straight for Heligoland, a small island in the North Sea that was mercifully free of blossoms and meadows. When he arrived, his face was so swollen that his landlady assumed he had been in a fight and promised to help him recover. From his second-floor room, Heisenberg had a breathtaking view of the village, the sand dunes, and the sea.

Heisenberg was disillusioned with physicists' attempts to tackle quantum theory and decided to start afresh. The only ingredients he used were properties of atoms that could be inferred from measurements, such as the colors of light that gases absorbed or emitted in the lab, otherwise known as their "atomic spectra."[12] By arranging the information in tables, Heisenberg started to develop a quantum theory that described the colors in terms of electrons jumping between different energy levels in an atom. They jumped up when they absorbed light and released it again when they fell back down.

The use of atomic spectra to understand the structure of the atom was a masterstroke. Shine light at an atom and it will absorb wavelengths that have just enough energy to boot electrons into higher-energy orbits. Later, when the electrons fall back down again, they release the energy, giving off the same color of light they absorbed. By studying which colors of light an atom absorbs and emits, you can work out the electronic orbits.

As Heisenberg's work progressed, the mechanics of the atom began to emerge from the shadows.[13] In his excitement, he made scores of mistakes and was filled with anxiety. "I was deeply alarmed," he wrote of his time in Heligoland. "I had a feeling that, through the surface of the atomic phenomena, I was looking at a strangely beautiful interior and felt almost giddy." Heisenberg completed his first calculation with his new technique at 3 A.M. one morning. Too excited to sleep, he left the house and walked down to the southern tip of Heligoland and climbed a rock that jutted out over the sea. There he waited for the sun to rise.

Back in Göttingen, Max Born looked over Heisenberg's mathematical formulas and realized his theory was right. There, with the help of a young theorist, Pascual Jordan, Born and Heisenberg developed Heisenberg's theory into what became known as "matrix mechanics," so-called because the terms used were written down in tables, or matrices.

Heisenberg's work was the first proper formulation of quantum mechanics, but a second theory soon arrived on the scene. In the run-up to Christmas 1925, the Austrian physicist Erwin Schrödinger, then at the University of Zurich, had his turn. He booked a few weeks in a picturesque, snow-capped lodge in the Austrian Alps and set off to work on his own quantum theory of atoms.[14] Schrödinger famously invited an old girlfriend to join him, leaving his wife, Anny, back in the city. Anny wasn't one to complain, though. To her, Schrödinger was beyond criticism. She also had a lover, her husband's close friend and university colleague, the mathematician Hermann Weyl.

Schrödinger used a completely different approach from Heisenberg's. His starting point was an idea put forward a year earlier by the French physicist Louis de Broglie, who imagined electrons to behave like waves.[15] Schrödinger worked steadily through the Christmas period and made no secret that he found the going tough. On December 27, he wrote to his friend Willy Wien, a Nobel Prize–winning physicist in Munich, to say: "At the moment, I am struggling with a new atomic theory. If only I knew more mathematics! I am very optimistic about this thing and expect that if I can only . . . solve it, it will be very beautiful."

By the time Schrödinger returned to Zurich, he had turned de Broglie's vague concept into a new version of quantum mechanics. Instead of Heisenberg's obscure matrices, Schrödinger's theory used a familiar wave-like equation. For the first time, it gave physicists a precise quantum formula they could use to describe particles in any atom or molecule. Whoever Schrödinger's companion was that Christmas, it's hard not to wonder if she got quite the romantic break she was expecting.

The existence of two versions of quantum mechanics made for an ugly start to the new era. There was no love lost between Heisenberg and Schrödinger. On seeing matrix mechanics, Schrödinger said that he was "discouraged, if not repelled" by it. Heisenberg's thoughts on Schrödinger's work—known as wave mechanics—were no more flattering: "The more I think about it . . . the more repulsive I find it," he said. The theories split the physics community into two defiant camps, but for no great reason. It turned out that, while the theories looked very different on paper, they were essentially the same. A mathematician could derive one from the other. History favored Schrödinger's theory and it was embraced by most physicists, not least because the mathematics he used was already so familiar to them.

For all its success, Schrödinger's equation had a major shortcoming. It did not seem to fit with the rules of special relativity, Einstein's extraordinarily disruptive theory of 1905. The failure was serious. It meant that if you tried to use Schrödinger's equation to describe highly energetic particles moving at or near the speed of light, the answers it gave were junk.

The job of marrying quantum mechanics with relativity—two of the most radical and powerful ideas physicists had—became one of the greatest challenges of the time. The breakthrough came from Paul Dirac, a profoundly gifted physicist who had studied engineering in the year above Thomas Higgs at Bristol University.[16] When Higgs moved to Newcastle, Dirac left for Cambridge University, where he switched to physics and immersed himself in the work of Einstein and the quantum pioneers, Heisenberg and Schrödinger.

At the end of 1927, at the age of twenty-five, Dirac wrote down an equation that many physicists regard as one of the most beautiful in the history of science. It marks Dirac's memorial at London's Westminster Abbey to this day. It reconciled quantum mechanics with relativity; explained the property of electrons known as spin, which causes magnetism; and even hinted at the existence of particles with the mysterious property of "negative energy." A particle with negative energy is identical to its more familiar counterpart, but has the opposite charge. A few years later, in 1932, Carl Anderson, a physicist at the California Institute of Technology, confirmed Dirac's perplexing prediction by discovering positively charged electrons, or "positrons," the first examples of what we now call "antimatter."[17]

Dirac, the failed engineer turned physicist, had pulled off one of the greatest coups in the history of quantum mechanics.[18] At the time, Freeman Dyson was a young boy living in Winchester in southern England. Later in life, he put Dirac on a pedestal, summing up those early years of discovery with these words: "The great papers of the other quantum pioneers were more ragged, less perfectly formed than Dirac's. His great discoveries were like exquisitely carved marble statues falling out of the sky, one after another. He seemed to be able to conjure the laws of nature from pure thought—it was this purity that made him unique."

Dirac's work brought within reach the modern description of all known matter in the universe. It enabled physicists to draw up a "quantum field theory" to describe the behavior of electrons and photons. Today, every known elementary particle is described by a broader quantum field theory, the Standard Model.

A year after Peter Higgs was born, the family moved to Birmingham and later back to Bristol, where they arrived a few days before the Good Friday bombing of 1941. It was at Cotham Grammar School in the city that Peter found his inspiration as a scientist.

Higgs's path into physics wasn't an obvious one. He loved languages and enjoyed mathematics and chemistry, but found physics lessons unexciting. Many of the younger and more dynamic teachers were away at war, leaving the stiffer old hands to teach Peter and his classmates. When it came to sitting for the exam for the School Certificate, Higgs picked up prizes for English, French, Latin, mathematics, and chemistry, but didn't excel at physics.

The daily routine began with morning assembly. Higgs took to standing at the back, where he whiled away the time reading the names immortalized on a large board that dominated one wall. They were the school's most prominent alumni. One name, Higgs noticed, appeared several times for different honors. It was Paul Dirac, the quantum theorist and only Nobel laureate on the board. The man leading the assembly was about to retire when Higgs arrived, but he had been the new headmaster when Dirac was in his last year as a pupil at the school.

Higgs was fascinated by Dirac, and it was he, more than any of the other founding fathers of quantum mechanics, who ignited Higgs's passion for physics. His English teacher urged him to read about science outside the classroom, especially the popular books by the Cambridge physicist Arthur Eddington, the cosmologist James Jeans, and Albert Einstein.

One of Eddington's tales had unfolded on May 29, 1919, ten years to the day before Higgs was born.[19] Eddington had cooked up a brilliant plan. He realized that nature provided a way of testing Einstein's theory of general relativity, which said massive objects created gravity by curving space around them. Eddington promptly set sail for the tiny island of Principe off the west coast of Africa and arrived in time to witness a total eclipse of the sun.

The sun has more mass than anything else in the solar system. If it didn't, the planets would orbit some other celestial body. Eddington's idea was that if the sun really curved space, the effect should reveal itself as a shift in the usual positions of the stars that became visible when the moon briefly blotted out the sun during an eclipse. The stars hadn't moved, of course, but the light coming from them would follow the curvature of space around the sun, making them look as though they had. Eddington's observations were published a year later and made headlines around the world as the first compelling evidence for Einstein's theory.

Higgs read voraciously. The stories Einstein, Eddington, and the others told gripped his imagination, though he was later dubious of some of Eddington's work. The physics he was learning in school was dull in comparison. All the good stuff, Higgs thought, was left off the syllabus.

The war years reshaped the intellectual landscape of science in a way that directly influenced the war's outcome. In the face of rising Nazism, Jewish scientists in Germany lost their jobs or left for posts in other countries. The anti-Semitic policies of the newly appointed chancellor of Germany, Adolph Hitler, caused a mass exit, effectively destroying the European powerhouses of physics in Berlin and Göttingen. Other scientists fled, too. The Einsteins moved to Princeton, New Jersey; Max Born went to Cambridge and then moved to Edinburgh University; Niels Bohr was smuggled to Britain in the bay of an RAF bomber and went on to America.[20] Erwin Schrödinger hoped to settle at Oxford University, but he was barred by college authorities who frowned on his wish to live with two women at once. Heisenberg, who was born in Würzburg, northern Bavaria, stayed on in Germany, where he later became director of the Kaiser Wilhelm Institute in Berlin.

Basic physics research was effectively put on hold for the war. On both sides of the conflict, scientists were diverted to technology projects to aid the war effort. Germany developed rocketry, though some of the biggest advances in that field had already been made by the American Robert Goddard. The British developed radar. Computing advanced on the back of code-breaking algorithms, while penicillin was mass-produced for the first time, saving countless lives on the front line.

Throughout the war, scientists made headway with an experimental technique called nuclear fission. Fission reactions release energy by splitting atoms of uranium and other materials. The Allied and Axis powers both knew that with the right expertise it was possible to create a chain reaction and release an enormous amount of energy from countless atoms in one devastating blast.

Paul Dirac spent the war years at Cambridge as the Lucasian Professor of Mathematics, the job that Isaac Newton had held more than 250 years earlier and that Stephen Hawking would assume 40 years later. Dirac worked briefly on confidential techniques to make weapons-grade uranium, which fed into the Manhattan Project, the U.S. atomic bomb effort led by Robert Oppenheimer at Los Alamos National Laboratory in New Mexico. Dirac mostly avoided military research, though, instead grappling with the challenges of being an academic at Cambridge. One of his wartime headaches was a brilliant if disruptive student named Freeman Dyson. Dyson amused himself by clambering onto the roofs of the buildings on campus during blackouts. But his favorite sport seemed to be bombarding Dirac with endless questions, on one occasion forcing him to end a lecture halfway through to find an answer to one of his queries.

The physics of fission was worked out in 1938, nearly a year before war broke out. German scientists, led by Otto Hahn in Berlin, blasted uranium with beams of neutrons and found that the atoms split into two lighter elements.[21] The reaction released energy, but also more neutrons. It dawned on the scientists that these neutrons could be used to split more uranium atoms, releasing yet more neutrons. Each step in the reaction would release more and more energy.

Hahn's work built on experiments the Italian physicist Enrico Fermi had done a few years earlier in Rome. Fermi discovered he could make radioactive elements in the laboratory by firing neutrons at uranium, but as far as he could tell, none of his experiments caused the uranium to fission—that is, split in two. At the time, most physicists thought neutrons were too weak to crack an atom into two lighter elements. Instead, the neutrons got lodged in the atoms and made them heavier. Fermi was awarded the Nobel Prize in 1938 for his work on "induced radioactivity." That same year, he and his Jewish wife, Laura, left Italy and headed for New York to escape Benito Mussolini's fascist regime.

Upon arriving in America, Fermi joined the Manhattan Project, where he was assigned to the inconspicuously named Chicago Metallurgical Laboratory and tasked with building a working nuclear reactor beneath an old sports stadium owned by the city's university. On December 2, 1942, James Conant, head of the National Defense Research Committee, took a call from the director of the Chicago laboratory, Arthur Compton. He had news of Fermi's progress. "The Italian navigator has just landed in the new world," Compton said. The oblique metaphor could mean only one thing: Fermi's reactor had achieved a chain reaction that released the energy of the atom. The era of nuclear power had begun.

The following year, Freeman Dyson left Cambridge to work in the Operational Research Section of RAF Bomber Command, which occupied a collection of red-brick buildings in a hilltop forest in the Buckinghamshire countryside. Dyson lodged in a nearby village and on the five-mile bicycle ride to work each morning would sometimes be passed by an RAF limousine whisking his boss, Sir Arthur "Bomber" Harris, to the office. Dyson's job was to make bombing raids safer for the pilots as well as more efficient at raising firestorms in order to inflict more damage on the older cities in Germany. The work, he later lamented, made him expert in "how to murder most economically another hundred thousand people."

Higgs was weeks from turning sixteen when news of Germany's surrender reached Britain. From the Cabinet room at Number 10, the prime minister, Winston Churchill, announced the ceasefire with trademark drama. His address was restrained, though. "We may allow ourselves a brief period of rejoicing; but let us not forget for a moment the toil and efforts that lie ahead. Japan, with all her treachery and greed, remains unsubdued," he said. Street parties broke out across Higgs's home city of Bristol and the rest of the country. When the pubs ran out of beer, the celebrations continued, with people dressed in red, white, and blue for the Union Jack dancing and singing around fires long into the night.

Three months later the war was brought to an end by a horrific display of technology. On the morning of August 6, 1945, the U.S. bomber *Enola Gay* released its 9,000-pound cargo over the port city of Hiroshima on Honshu, the largest island of Japan. It fell for 43 seconds and detonated 2,000 feet above the city. Inside the bomb, a small conventional charge near the tail fins exploded first. The blast propelled a pellet of enriched uranium down a metal barrel and into another lump of uranium near the nose of the bomb. When the two met, an initiator injected neutrons into the uranium, which triggered a chain reaction. The first neutrons split a few uranium nuclei, which released more neutrons in the process. These split other nuclei and released even more neutrons. As the reaction gathered momentum, huge amounts of energy were released until the bomb tore itself apart in the sky. The *Enola Gay* was nearly 12 miles away when it was rocked by the blast. Moments later a second shockwave struck, a reflection of the first off the ground, and the crew turned to look back. The city was obscured by a towering mushroom cloud.

News of the attack was broken later that day to German scientists who had been captured and interned at Farm Hall on the outskirts of Cambridge earlier in the year. Among them were Otto Hahn and Werner Heisenberg. Hahn was inconsolable. He felt responsible for the deaths of hundreds of thousands of people. He confessed to having considered suicide after realizing the potential of his work on nuclear fission. Transcripts from the bugged conversations reveal horror and disbelief among the other scientists whose work made the bomb possible.[22]

Werner Heisenberg listened in disbelief. He had thought it would take years to build the bomb and was convinced the report was propaganda. But he had made basic errors in calculating the critical mass of uranium that was needed, thus overestimating the difficulty involved in making the bomb work. The Hiroshima bomb released the energy of only a kilo, or about 2 percent, of its uranium, but it was enough to destroy several square miles of the city and kill tens of thousands of people.

The discussions at Farm Hall reveal what went through the scientists' minds as the consequences of their work sunk in. Otto Hahn tried to comfort another German scientist, Walter Gerlach, asking: "Are you upset because we did not make the uranium bomb? I thank God on my bended knees that we did not make the uranium bomb." Gerlach responded: "You cannot prevent its development. I was afraid to think of the bomb, but I did think of it as a thing of the future, and that the man who could threaten the use of the bomb would be able to achieve anything."

In an interview for a television documentary in 1980, Freeman Dyson was candid about the mixed feelings he had toward the bomb:

> I feel it myself, the glitter of nuclear weapons. It is irresistible if you come to them as a scientist. To feel it's there in your hands. To release the energy that fuels the stars. To let it do your bidding. And to perform these miracles, to lift a million tons of rock into the sky, it is something that gives people an illusion of illimitable power, and it is in some ways responsible for all our troubles. I would say this is what you might call "technical arrogance" that overcomes people when they see what they can do with their minds.[23]

The bombing of Hiroshima and subsequently Nagasaki prompted a flurry of physics lectures on nuclear fission and the science of the atomic bomb at Bristol University. Peter Higgs went to them. Some students were spellbound by the impact their subject had had on the course of history. Others were repulsed. "The bombs on Japan either drew people into theoretical physics or made them leave in their droves," Higgs said later. "That's when I began to learn about what was going on. After that, I decided to avoid anything that was to do with weapons."

At the end of the war physics was viewed in a new light. The conflict demonstrated beyond doubt that discoveries in the field would influence the course of events at every turn. The cost of being slow, of failing to make a discovery or exploit it, was all too clear. As nations rebuilt themselves, physics was put on a pedestal. It was time for a new generation of physicists to go further into the quantum world and discover the origin of mass.

3

Seventy-Nine Lines

At five o'clock one May afternoon in 1950, Peter Higgs got up in front of the Maxwell Society at King's College London and started an argument that never ended.[1] Higgs was twenty years old and the student president of the society, which met every week to discuss science outside university lectures. Higgs questioned whether scientists could ever really know the laws of nature. He wasn't so sure they could.

A scientist's impression of the world, like anybody else's, is built on a collection of lifetime experiences. Those experiences form when our senses tell our minds what is going on in the world. We see things happen, hear the noises they make, feel their movement, smell their odors, taste them. Our brains take all this sensory data and from it construct a view of the world. Our own very personal view of reality.

Science is the art of finding patterns in reality. How do shadows fall as the sun moves in the sky? What does light do when it shines through a prism? What direction do you go in when you fall over? Scientific experiments are designed to elicit these patterns. If they produce the same result time and again, the pattern might be important. It might reveal a law of nature.

Higgs's concern was simple. How can scientists be sure that the observations they make are real? Are we sure our senses give a faithful account of the physical world? Does our brain do a flawless job of stitching our experiences into an accurate picture of the world? To believe so was an act of faith, not logic, said Higgs. Our minds might fail us. We might dream we did an experiment but think it was real. We might be deluded and see things that didn't happen. What might seem like a pattern in reality, a law of nature, might be nothing more than a trick of the imagination.

The minutes of the meeting that May afternoon are recorded in neat black ink on the yellowing pages of a burgundy book held in the King's College archives in London. The entry for that day begins by describing Higgs's argument as it unfolds and then turns to the reaction of the staff and students in the room. "This point aroused considerable controversy," the notes say.

Higgs's question was a familiar one for philosophers. In the earliest days of modern science, the seventeenth-century French philosopher René Descartes was working on two great problems: what can we know and how can we know it. Descartes framed his argument by imagining there was a demon intent on fooling him with evil mind tricks. If that were the case, what could he still know was true for sure? What couldn't the demon fool him into believing? Descartes whittled it down to a singular answer. The only thing we can know for sure is that if we are thinking, then we exist: *cogito ergo sum*.[2] I think, therefore I am.

If Descartes had stopped there, he would have left us with a bleak and solitary world. But he didn't. Descartes turned the problem around. What if there was a benevolent God, he said, one that gave us senses that were true and brains to match? We could be quite sure that our view of reality was accurate. If God is good, we can trust our senses and be happy that the world we see is the world that exists. Higgs didn't look to God for a way out of the dilemma. He said he'd feel a lot happier about believing a scientific result if scores of people did the same experiment independently and got exactly the same answers. Higgs's friend Michael Fisher was in the audience and up in arms. He said he was searching for answers in his own world and he could do it alone. "I have to rely on my senses!" he said. The meeting was adjourned in deadlock.

The Maxwell Society was named in honor of James Clerk Maxwell, who had written down his theory of light at King's College some eighty years before Higgs had arrived. The society met in room 2C, the first room in Britain to be lit by gaslight and the room that the nuclear scientist Alan Nunn May was lecturing in when he was arrested for being a Russian spy in 1946. A month before the bombings of Hiroshima and Nagasaki, Nunn May had passed a small quantity of weapons-grade uranium to his Soviet handler, and he later forwarded details of the Japan bombings. He received $200 and a bottle of whisky in return.

More often than not, the Maxwell Society brought in guest speakers. For Higgs these were an unmissable part of college life. Arthur C. Clarke, a recent graduate of the college, came by and talked about interplanetary travel.[3] Sir Edward Appleton, who received the Nobel Prize for discovering an upper layer of the atmosphere, called the "ionosphere," by bouncing radio waves off of it, discussed the chances of picking up signals from extraterrestrials. Charles Coulson, the college's first professor of theoretical physics, talked about the future of physics, concluding that scientists would know it all in ten or twenty years, unless something unexpected turned up, such as telepathy.

The Higgs family had moved to London in 1946. Peter had joined King's College to study physics, but he soon realized he was not a natural experimenter. He would wrestle with equipment for ages and still not get it to work. On one occasion, he tried to repeat the classic "Millikan" experiment from 1909 that involves using an electric field to make charged oil droplets hover in midair, but he couldn't get a single damned drop to hold still.

The year Higgs started at King's College, Freeman Dyson fulfilled a dream. He was tired of Britain's postwar languishing and hankered for the brighter optimism of America. At the age of twenty-three, he made the break and embarked on a voyage to New York;

he'd been hired to work at Cornell University under Hans Bethe, who had run the theoretical division of the Manhattan Project.

At Cornell, Dyson quickly found himself immersed in a nasty problem that was threatening to scupper progress in quantum physics. The problem involved the need for a theory that described how atoms and electrons absorbed and emitted light. The theory was part of the grand and ongoing effort to explain all the different kinds of particles and their interactions in terms of quantum mechanics.

At the heart of the problem was a glitch in the quantum field theory that physicists had drawn up to describe photons and electrons. The theory, known as "quantum electrodynamics," went haywire under certain circumstances.

Robert Oppenheimer had drawn attention to the flaw as a twenty-six-year-old in 1930, but when the war intervened the problem was put to one side. Oppenheimer had been using quantum field theory to work out what happened if an electron emitted a particle of light but quickly reabsorbed it again. It's the electronic equivalent of throwing a tennis ball in the air and catching it again. You lose energy when you throw it, but gain energy when you catch the ball on the way back down.

The amount of energy used in throwing and catching a tennis ball is far too small to cause a human any trouble, but emitting and absorbing a particle of light can play havoc with an electron. Nature doesn't impose an upper limit on how much energy the particle of light might have, and electrons could emit countless of these "virtual photons" all the time. Oppenheimer worked out that these fleeting photons could cause infinite changes in the energy of an atom. That was impossible, which could mean only one thing: there was something badly wrong with the theory. It was fine as a rough guide, but nothing more.

One of the physicists Dyson fell in with at Cornell was Richard Feynman, a self-assured and brilliant New Yorker who was determined to rescue the beleaguered quantum field theory.[4] In 1947, Feynman realized that, instead of worrying about each and every particle of light that popped into existence around an electron, it was better to pull back and view the particles more as a cloud of energy. Do this and you could redefine the mass and the charge of the electron. When Feynman worked through the mathematics of taking the effects of the photon cloud into consideration, the troublesome infinities that threatened to derail the theory vanished.

Feynman's work was crucial to what became known as "renormalization." It was the breakthrough physicists needed to put quantum electrodynamics on a firm footing. It meant the theory worked not just for slow-moving, low-energy particles, but also at high energies when particles were zipping around at or near the speed of light. Feynman's work should have sparked celebrations in university physics departments around the world, but there was a complicating factor: two other physicists had come up with radically different ways of solving the crisis. One was Julian Schwinger at Harvard University; the other was Sin-Itiro Tomonaga at Tokyo University. Tomonaga had cracked the problem during the war, but the news took years to reach the West. The situation had all the ugliness of the late 1920s, when Heisenberg and Schrödinger had developed two competing formulations of quantum mechanics.

In the summer of 1948, Dyson and Feynman took off on a road trip from New York to Albuquerque in New Mexico. Along the way they talked physics, picked up hitchhikers, spent a night in a brothel (the local hotels were full and they sought only shelter), and, upon arrival at their destination, ran into trouble with the police for speeding. In Feynman's case, a girl was involved.[5] While he headed off to find her, Dyson carried on alone.

Dyson went to look for America aboard the dilapidated Greyhound buses that ran between desolate terminals on the rougher edges of the cities. After a couple of weeks traveling, he found himself in California and boarded a Greyhound for the long trip home. Staring idly out of the window somewhere in the middle of Nebraska, Dyson had a sudden thought. Physics had not been on his mind for weeks, but the realization burst into his consciousness like an explosion. He could see the work of Feynman, Schwinger, and Tomonaga with a fresh clarity. They looked so unalike, but Dyson understood that they effectively said the same thing, albeit in radically different ways. He didn't have a pen or any paper on which to write down his thoughts, but he didn't need them. He had seen in his mind how to combine the three theories into one. Dyson checked his idea through when he got back to Cornell and saw it checked out. He published his work in 1949, causing a sensation.

Science progresses in fits and starts. It lurches from one problem to another, and the problems are never-ending. Dyson's work topped an edifice that had its foundations in Maxwell's field theory of electromagnetism and had grown with each breakthrough in physics. With Dyson's work in place, scientists finally had a fully functioning quantum field theory that explained the behavior of some of the most important particles in the universe: electrons and photons. Feynman was so impressed with quantum electrodynamics (QED) that he called it the jewel of physics.

The hard-won success of quantum electrodynamics set the course of theoretical particle physics for the rest of the century. Scientists looked to expand quantum field theory to explain the behavior of other particles, such as the subatomic constituents inside atomic nuclei. Quantum field theories assume the world is made up of a variety of different fields. The quanta of these fields are the particles that make up matter or transmit forces between them. The electron is the quantum of the electron field. Particles of light, or photons, are quanta of the electromagnetic field and carry the electromagnetic force.

Back in London, Peter Higgs was in the final year toward earning his degree. His often hopeless attempts to do experiments convinced him that his future lay in the theoretical aspects of physics. His timing was fortunate: King's College had just started up a course in theoretical physics, and Higgs became the first and only student in his year to take it. At the end of the academic year, his professors realized they had a problem. They needed to give Higgs an exam, but no one knew what questions to ask him: they hadn't had to worry about a theoretical physics exam before. One of the professors hit on a bright idea. They found a physics paper that had just been published in one of the top academic journals. Knowing Higgs could hardly have seen it, let alone read it, they reworded the paper into a puzzle for him to solve.

More than fifty years later, Michael Fisher, now a distinguished professor at the University of Maryland, had forgotten his philosophical argument with Higgs at the Maxwell Society all those years earlier, but the college mates had become lifelong friends. Fisher does remember the fuss over Higgs's final year examination, though. When Peter handed his paper in, the professor who had determined the question looked at it in disbelief. Higgs had not only got the answer right but had gone one better. "His solution was better than the original!" Fisher said.

Higgs graduated, but that meant he had to decide what to do next. He desperately wanted to do a Ph.D. on quantum field theory and knew that Paul Dirac was still at one of the best groups in the world at Cambridge University. There were two problems, though: one was Dirac, the other was Cambridge. Higgs was wary of the perceived privilege of Oxbridge and those who went there, and Dirac was notoriously taciturn, to the point that scholars have speculated that he might have been autistic. It made many of his relationships awkward, and he didn't often take on Ph.D. students. When he did, overseeing their work was a chore, and he rarely expressed much interest in their progress. One young physicist, Dennis Sciama, who in later life supervised the British cosmologists Stephen Hawking and Sir Martin Rees, was briefly Dirac's Ph.D. student and experienced the effects of his temperament after having a bright idea about the cosmos. Sciama hurried along to his supervisor's office and knocked on the door. When Sciama was called in, he said: "Professor Dirac, I've just thought of a way of relating the formation of stars to cosmological questions, shall I tell you about it?" Dirac issued a simple "No," leaving Sciama with little choice other than to walk back out.

There was a more serious problem that influenced Higgs's future. He talked his options over with Charles Coulson, who warned him that quantum field theory was in a mess. "It's the kind of field where you either achieve nothing or win a Nobel prize," Coulson said. Higgs didn't know it at the time, but Coulson's fears were out of date: the problem he had in mind was the one that had been sorted out at least a year earlier by Freeman Dyson, Richard Feynman, and the others. Higgs decided to play it safe and stayed on at King's College. His Ph.D. studies focused on theoretical problems in chemistry that were important for understanding the structure of molecules.

When young scientists finish their Ph.D.s, they usually spend time in what can be an academic no-man's-land of postdoctorate fellowships.[6] These are typically one- or two-year posts, and they serve universities well by keeping them supplied with cheap labor. Higgs spent several years on the postgraduate circuit. There were two years at the physics department in Edinburgh, a city he had fallen in love with when he hitchhiked there during the international festival in 1949. He took on other posts afterward, including one at Imperial College in London and another at University College London (UCL), where he lectured in mathematics.

Higgs arrived at UCL in 1960 and temporarily took on a second job as secretary of the scientific group of the Campaign for Nuclear Disarmament (CND). The organization had been set up two years earlier by eminent figures on the political left who wanted a unilateral ban on nuclear weapons. The first marches on Aldermaston, Britain's nuclear bomb facility, had seen enormous turnouts, and the CND symbol seemed to be everywhere.

Higgs was responsible for arranging talks from scientists who backed the campaign. In March that year, he hoped to invite the American peace activist and scientist Linus Pauling, who had won a Nobel Prize for work on chemical bonds in 1954. The prominent Oxford-based ophthalmologist Antoinette Pirie wrote to Pauling urging him to let Higgs know if he could come to London for "a [CND] soirée or cocktail party, or perhaps to talk to a meeting of

CND." The letter, written on March 25, 1960, described Pirie's assessment of the nuclear issue in Britain at the time: "The official opinion in Britain is swinging away from nuclear weapons but for all the wrong reasons, e.g. expense, but a lot of political parties are still committed to reliance on America's deterrent." The letter ended on an optimistic note: "Perhaps the new Russian acceptance of a partial ban with control will evoke a positive response from Mr McMillan [*sic*] or Mr Eisenhower. Even to go a little way in the right direction would mean so much." Both the British prime minister, Harold Macmillan, and the U.S. president, Dwight D. Eisenhower, strengthened their nuclear deterrents throughout the Cold War.

The letter reached Pauling at his home in Pasadena, California, but was lost amid a pile of papers. Higgs received a reply, with profuse apologies, three months later. Pauling's letter emphasized how much pressure he was under at the time. He had been ordered to appear before a security hearing of the Senate Judiciary Committee that summer. The committee had asked him to bring along the names of all the people who had helped gather signatures for a petition Pauling had organized that urged the United Nations to introduce an international agreement banning nuclear weapons testing. Pauling made clear that he had no intention of complying. He signed off saying he dearly hoped he hadn't left it too late to come to visit the CND in London. Two years later, Pauling was awarded the Nobel Peace Prize, making him one of only two people to win two different kinds of Nobels. The other was Marie Curie, who won the physics prize in 1903 for her research on radioactivity and the chemistry prize in 1911 for discovering the radioactive elements radium and polonium.

In the autumn of 1960, Higgs got the job he had long hoped for. Nicholas Kemmer, one of Dirac's colleagues at Cambridge, had left for Edinburgh University, where he took over the professorship recently vacated by Max Born. Kemmer was looking for a physics lecturer, and Higgs was the ideal candidate. Kemmer could not have been more different from Dirac. Chatty and easy-going, he joked with his protégés that he'd lost touch with cutting-edge physics many moons ago. Kemmer was also active in the CND and soon pushed piles of his campaign work Higgs's way, including the organization of the regular CND staff meetings.

Higgs was in no mood to complain. His involvement with CND transformed his social life. It was at the staff club in his first year at Edinburgh that Higgs got to chatting with a twenty-four-year-old who was drinking with his CND friends and had moved to the city from Urbana, Illinois, to study speech development. Her name was Jody Williamson. The two got on like a house on fire.

At the physics department, Higgs took charge of the academic journals that landed in the secretary's office each week. He flicked through them, marked the date on the top, and put them out on display for the other researchers. The one perk of the job was that he got to see the journals before anyone else.

In the spring of 1961, Higgs was leafing through one of the newly arrived journals when an article caught his eye. A Japanese American physicist, Yoichiro Nambu, at the University of Chicago had written a paper that drew on the theory of superconductors to explain how elementary particles may have acquired their masses.[7] Nambu had worked with Einstein before moving to Chicago and had a huge reputation. A colleague once said of him that he was often so far ahead of the game they found him unintelligible.[8]

A normal conductor, such as a copper wire, conducts electricity because of the way its atoms sit together. The atoms form a lattice in which the electrons orbiting one copper atom overlap with those of the atom next door. It is effectively a lattice of positive copper ions in a sea of almost "free" electrons. Since the electrons can roam around so easily, copper wire conducts electricity extremely well. The electrons move through it like water down a garden hose.

The conductivity of copper, like that of other metals, depends on its temperature. Heat a lump of metal and it becomes a worse conductor, because the ion lattice vibrates and obstructs the flow of electrons. When it is cooled down, the vibrations are smaller and electricity can flow more easily. A normal metal never becomes a perfect conductor, though, because even at absolute zero, or −273.15 degrees Celsius, electrons are still held up by defects and impurities in the lattice.

A superconductor is different from a normal conductor in a crucial way. If you cool a superconductor down, then at some point below −130 degrees Celsius it will suddenly lose all of its electrical resistance. The strange behavior of these materials led scientists to dream of ultra-efficient power grids that theoretically would use superconducting wires to ferry electricity around countries and even between them without losing any energy.

In the late 1950s, scientists worked out what makes superconductors live up to their name. When a superconductor is cooled below a critical temperature, the electrons inside get together in pairs. The effect is extraordinary. When electrons pair up in superconductors, they act as though the lattice isn't there. Scientists say they behave like a "superfluid," a substance that moves without losing energy. As long as the superconductor stays below its critical temperature, electricity will pass through it without the slightest resistance.[9]

The internal goings-on that cause a superconductor to suddenly lose all resistance is an example of what physicists call a "broken symmetry."[10] Nambu's conceptual breakthrough was to wonder if some other kind of broken symmetry, one that happened across the universe, might have made massless particles massive. His paper roughed out how such a thing could happen, giving mass to protons, neutrons, and a few other particles. The work didn't prove anything, but it planted a seed in physicists' minds, Higgs's included, that a broken symmetry might be the key to the origin of mass.

It is hard to overstate the role that symmetry has played in the history of physics. Ever since the days of Galileo, physicists have looked to symmetry as a guide to understanding the laws of nature. By symmetry, physicists mean properties of nature that stay the same under different circumstances.

You can see symmetry at work all around you. A billiard ball that was a solid color (without even a number on it) would look the same from every angle because it would be completely symmetrical. Spin it around like a top and its appearance would stay the same. This is an example of what is called "rotational symmetry." There are countless other kinds of symmetry though. Pick the ball up and put it down on the table in a different spot and it still looks the same: an example of "spatial" or "translational symmetry." Wander off and come back to the ball in ten minutes and it looks the same still: that's "temporal symmetry." What does any of this tell us? That the appearance of the ball is not influenced by its position in space or time.

Symmetries are so deeply rooted in our experience that we take them for granted. To physicists, symmetry is a tool for understanding how the world works. If you know the symmetry of an object or a process in nature, you are well on your way to understanding it. Suppose you tell some physicists you are holding something perfectly symmetrical in your left hand. They will probably guess you are holding a sphere of some kind. Now suppose you tell them that in your right hand you have something that is completely symmetrical around the vertical axis, but when you spin it around any of its horizontal axes it only looks the same after every full rotation. They might guess you are holding a billiard cue. Look down at the chalked end, and it looks the same no matter how much you spin it. But spin the cue end over end, or sideways in the vertical plane, and it only looks the same when the chalked tip returns to its original position. Knowing the symmetry of an object can help us understand what it looks like. Incidentally, our physicists might have just as correctly guessed you were holding a pencil, an ice-cream cone, or even a sombrero.

In 2008, Yoichiro Nambu won the Nobel Prize in Physics for his work on broken symmetries. When Professor Lars Brink, a member of the Royal Swedish Academy of Sciences, presented the prize, he began with a simple sentence: "The Earth is round." He went on to say not only how humans see significance in symmetries, but also how important symmetries are in determining the laws of physics. The Earth, like every other planet, is round because gravity is symmetrical. It pulls equally in all directions, that is, toward the center of the mass that is generating it.

The Earth helps us understand that symmetrical laws don't have to shape the world in a symmetrical way. The laws of physics that govern how big our planet is and how it spins through space are symmetrical, but you don't have to look too closely to see that the Earth isn't a perfect sphere. Because the planet spins, it is fatter around the equator. The movement of continental plates has given us awe-inspiring mountain ranges. The Earth is proof that even when the laws of physics are symmetrical, the outcome of those laws needn't be. The natural world hides the symmetry of the laws that govern it.

Every time you climb out of bed you witness a kind of broken symmetry. The Earth's gravitational field breaks the directional symmetry of our world. Gravity defines which direction is down, and, once that happens, up, left, and right fall into place. Without gravity, which way is "down" is purely arbitrary.

Deviations from symmetry can sometimes be more interesting than symmetries themselves. Take a look in the mirror. If you are good-looking, and I'm guessing you are, you'll notice how symmetrical your face is. You don't have one eye drastically lower than the other, and your ears are more or less level and jut out about the same amount on each side. Your nose and mouth evenly bridge a line that runs right between your eyes to the center of your chin. Let's not worry about what you've done with your hair today.

Many scientists think that facial symmetry is interpreted as beauty because we see it as a sign of healthy genetic development. The theory has a lot of evidence to back it up. Biological glitches that stymie the expression of certain genes can give rise to medical conditions that are characterized by tell-tale facial asymmetries. Certain distinctive facial asymmetries are a sign of a genetic condition called Fragile X syndrome, the most common inherited mental impairment. In biology as well as in physics, symmetry, or the lack of it, can give us profound insights into processes that are hidden from view.

Nambu said a spontaneously broken symmetry lay at the heart of particle mass. You can see spontaneous broken symmetry at work if you stand a pen on one end and let it go. When the pen falls over, it will be pointing in one direction or another. The pen goes from standing up in a symmetrical position to lying down in an asymmetrical one. The loss of symmetry is inevitable: the pen gives in to the tug of Earth's gravitational field and falls over. Nambu's work suggested that the universe began in a symmetrical stage in which all particles were massless. Then, thanks to a new kind of field, the symmetry was broken and some particles suddenly found they had mass.

Nambu's idea was compelling, but it had a flaw that Nambu himself was aware of. The British physicist Jeffrey Goldstone pointed out that the kind of symmetry-breaking Nambu proposed came with an inevitable side order of massless particles.[11] That is, new particles would have been created in the process. If these unknown particles existed, they would easily be made in nature and would pour out of the sun and other stars. We would see them everywhere. The inescapable fact that we don't see them suggested Nambu's theory was wrong.

Higgs wasn't the only one who had seen something profound in Nambu's research despite its obvious flaw. At Cornell University, two physicists, Robert Brout and François Englert, were convinced Nambu's paper had deep implications for particle physics. The two had met years earlier when Brout, who was already at Cornell, needed an assistant. Brout asked his friend Pierre Aigrain, a prominent European physicist, to recommend someone. Soon after, Englert, an engineer-turned-physicist who had worked for Aigrain at the Free University of Brussels, was on a plane to America.

Englert was anxious about the trip. He hadn't met Brout before and didn't know anyone in America. He needn't have worried, though. When he emerged from customs in New York, Brout was there. He suggested the two go for a drink and they headed for the bars of Ithaca, the town where Cornell is based. One drink followed another and conversation drifted away from physics to everything else going on in their lives. In time, the two became as close as brothers, happily finishing each other's sentences and lampooning their clumsy attempts to speak each other's mother tongues.

Not long after Nambu's paper came out, Englert grew restless. He missed the excitement of city life and thought about returning to Brussels. He turned down a faculty position at Cornell and headed back to Belgium in 1962. Brout, who had spent time in Belgium himself, followed soon after, and both took up academic jobs at the Free University. There the two started combing over the details of Nambu's work.

In Cambridge, Massachusetts, a third group of physicists was circling the same problem that Higgs, Brout, and Englert were already working on. None of the sets of researchers knew what the others were up to or even that they were in competition for what became one of the greatest prizes in modern physics. As undergraduates, Gerry Guralnik and Dick Hagen were inseparable. Guralnik was a student at Harvard, while Hagen went to the Massachusetts Institute of Technology, but the two universities shared lectures that both attended. The two students were awestruck by Julian Schwinger, the Harvard physicist whose work had shaped quantum electrodynamics in the 1940s. When he arrived for a lecture, he started by writing down equations at the top left corner of the board and kept writing until he reached the bottom right, at which point he stopped and quietly strolled out of the room. His lectures seemed impenetrable at first, but for students who persevered, the man's genius was clear.

In 1964, Guralnik moved to London for a fellowship at Imperial College, where the brilliant Pakistani physicist Abdus Salam presided over a group of theorists who were arguably the world experts on the physics of broken symmetries. Guralnik and his wife, Susan, moved into a modest flat in Hampstead in north London. While Susan studied for a Ph.D. in history, Guralnik got to know his new colleagues. One was Tom Kibble, a tall, elf-like fellow with a formidable intellect. Kibble introduced Guralnik to the joys of Imperial College cuisine, taking him for lunches of vile hardboiled eggs and desserts drenched in something resembling custard.

Guralnik and Kibble began working on the idea that particle masses might be a result of broken symmetry. Guralnik knew they would make better progress with Hagen on board, so he made arrangements for him to come to London for an extended visit. Hagen moved in with Guralnik and his wife, making the Hampstead flat a microcosm of American familiarity in the heart of England.

University life wasn't much different from academia back home, but Guralnik and Hagen were unprepared for the world beyond the ivory towers. The atmosphere at Imperial was more formal than Guralnik was used to, so he took the subway into London to buy a new suit. At the tailor's, Guralnik announced he wanted a vest and two pairs of pants, a request that raised eyebrows among the staff. He later realized he had ordered what amounted to a tweed undersuit.

The Hampstead flat had no heating, and it was a bitterly cold start to the year in London. To warm the place up, Guralnik bought an electric heater and kept it on full blast. One day, Guralnik heard a yelping sound coming from the living room. He ran in to find Hagen jumping around with smoke rising from his trousers. Hagen had come home so frozen to the bone that he'd stood right in front of the fire to warm himself up. At that moment, Guralnik understood why so many of the miniskirt-wearing girls on the tube had red stripes on their legs. They got so cold they stood too close to their heaters.

There is a dying tradition in science that arguments play out in the pages of academic journals. When a physicist publishes a paper that others don't agree with, they send their criticisms to the editors and ask them to print it. The original author then gets the right to reply. It's a civilized—if not particularly speedy—way of discussing the merits of scientific work. In the spring of 1964, one such disagreement broke out in the pages of the American physics journal *Physical Review Letters*. A year earlier, Philip Anderson, a physicist at

Bell Laboratories in New Jersey, had pointed out that the massless particles that seemed to be a big problem for Nambu's theory probably weren't a show-stopper after all. Anderson, who had been awarded the Nobel Prize in 1977 for his work on the structure of materials, said that massless particles also appeared in superconductors but immediately became heavy because of the way they interacted. He thought Nambu was on the right track. The argument began when Ben Lee and Abraham Klein, physicists in Pennsylvania, published some ideas in the journal they thought might improve on Nambu's work. Their two cents' worth was quickly followed by a letter from another physicist, Wally Gilbert of Harvard, who trashed their contribution.[12] On reading Gilbert's letter, Higgs was deflated. It meant his own work was likely to be wrong too.

Higgs was prepared for a downbeat weekend in Edinburgh, but as he mulled over Gilbert's letter he was struck by a thought. Gilbert had missed something. There was a loophole in his argument. Higgs knew a mathematical trick, used by Julian Schwinger to formulate quantum electrodynamics, that avoided the flaws Gilbert had spotted. If Higgs was right, it showed how particles could acquire mass without running into the problems that plagued Nambu's theory.

That Monday morning in July 1964, Higgs headed to the office and got to work. In seventy-nine lines of text and equations, he described the flaw in Gilbert's argument. He typed it out and sent it to CERN, the European nuclear research organization near Geneva, which was home to the editor of a journal called *Physics Letters*. His letter arrived at the end of July.

Not long after Higgs sent his first paper for publication, a letter arrived from America. It was from Wally Gilbert, who had read a preprint of the article. He was polite but firmly pointed out that Higgs had it all wrong. Higgs never got around to replying, and it only dawned on him ten years later that a small omission in the paper had caused Gilbert to misunderstand it. "I didn't spot it at the time. I was so excited I wrote down what I'd already worked out really quickly. Far too quickly," Higgs recalls.

Higgs completed his second paper a week later and sent it off to CERN. The editors of *Physics Letters* asked independent referees to review the paper before they published it, a standard practice at all academic journals. One morning, a letter arrived from CERN. The journal's editor, Jacques Prentki, said Higgs should do some more work on his theory and submit it to another journal, suggesting one in Italy that didn't bother with referees. Higgs later found out that Prentki thought his paper had "no relevance to particle physics."[13]

Higgs was dismayed and not a little indignant. He realized he must have sorely undersold his work. After rereading the paper, he tacked on a couple of new paragraphs at the end. In the penultimate sentence, Higgs pointed out that the theory came with its own signature particle. It was the sentence that introduced the Higgs boson. Higgs next followed Prentki's advice and sent his revised paper to another journal, but, instead of posting it to Italy, he sent it to *Physical Review Letters* in the United States. It was the main rival to the CERN-based journal.

It was September before Higgs heard back. His paper had been accepted but with one proviso. The referee wanted Higgs to mention another paper that was published the day his own had arrived at the journal's offices.[14] The paper had been written by two physicists in Brussels, Brout and Englert. Using a different approach, the two had arrived at a similar theory to Higgs's. They beat him into print by seven weeks. One major difference between the articles was that Higgs's paper made specific reference to the new particle that later came to bear his name.

In Brussels, Brout and Englert knew nothing of Higgs's work. They headed out to celebrate and wound up in a beautiful seventeenth-century terrace café overlooking a park in the city. Fresh from the excitement of their breakthrough, they toasted each other's health. Decades later, Brout recalled what was going through his mind. "For the first time in my life I felt what it might be like to be a great physicist," he said.

Meanwhile, at Imperial College in London, Guralnik, Hagen, and Kibble had worked out their own version of the theory that explained how particles gained mass through symmetry-breaking and had written it up for publication. Guralnik and Hagen had given it a final read and had it in an envelope, ready to send to *Physical Review Letters*, when Kibble walked through the door waving three papers. Two were written by Higgs, the other by Brout and Englert. They had been held up in the mail or lost when they arrived at the college.

A quick read-through of the new papers showed they covered achingly similar ground, but the three felt their own paper was more complete. They decided to make a few hasty additions to mention the work of Higgs, Brout, and Englert, and sent it off for publication. Their article appeared in the journal on November 16, 1964.

The three groups described a new kind of field that lives in the vacuum of space.[15] Their theories said that, when the field switched on, it gave mass to some particles but left others alone.[16] The process they were proposing has striking similarities to what physicists now know goes on inside a superconductor. The rearrangement of electrons in a superconductor breaks the electromagnetic symmetry, and this has an unusual consequence. When a photon—a massless particle of light—meets a superconductor, it immediately becomes heavy. Inside a superconductor, the broken symmetry makes photons massive. Physicists call it the Meissner effect.

In the summer of 1965, Werner Heisenberg organized a small meeting in the picturesque town of Feldafing on the shores of Lake Starnberg, outside Munich. The meeting drew a lot of the old-timers in physics, including Edward Teller, who had worked alongside Robert Oppenheimer on the Manhattan Project and pushed through the development of the hydrogen bomb during the early years of the Cold War.

Guralnik and Hagen liked the sound of Heisenberg's conference, and it was the perfect excuse to see more of Europe. They decided to make it a road trip. Their first goal was to get to France. From the border, they headed to Paris, where Hagen picked up a cheap Renault 8. After dipping a tentative toe into Parisian cuisine and learning what an artichoke was, the two drove off to Bavaria.

Both were scheduled to speak at the conference. Guralnik had already given some lectures on the theory in Europe and been taken aback at the frosty reception he received. The theory was met with "almost uniform disbelief," he told me years later. Even so, he wasn't prepared for what happened in Feldafing. Of all the people who criticized the theory, Heisenberg was the most vocal. The theory, he said, was "junk."

That August, Peter and Jody left Scotland for Higgs's year-long sabbatical at the University of North Carolina in Chapel Hill. The papers Higgs had published described how nature might give mass to certain particles. But they were only the beginning. It was clear that the physics community was far from convinced by the theory. With no hard evidence that it was right, the theory wasn't much use. It might be a clever trick that nature ignored. The search for evidence was soon at the forefront of physicists' minds. A new race was on.

4

The Enchanted Prince

Science fails when scientists march in the same direction.[1] It becomes an exercise in flogging the old instead of rushing forward with the new. For a true revolution you need contrarians, not conformists. And so it was with the six men who stumbled on the origin of mass. When Peter Higgs began work on the theory at Edinburgh University, he was considered an outsider. He stood out because he didn't do what everyone else was doing.

In the early 1960s, physicists were abandoning quantum field theory in droves.[2] This was the kind of theory that Freeman Dyson and others had worked on nearly two decades earlier, successfully describing how atoms emitted and absorbed particles of light. The disaffection set in when physicists failed to expand quantum field theory to explain other forces and particles. The particle physics community became deeply jaded. They decided quantum field theory was a one-trick pony.

In all but a handful of universities, physicists dropped quantum field theory and put their faith in an altogether different kind of calculation that emerged as a hopeful contender to push particle physics forward. The theory was known as S-matrix theory.[3] It was a mathematical framework that sought to explain the behavior of particles by comparing them before and after they interacted or bounced off one another.

Higgs took a dim view of S-matrix. In S-matrix work, one formulated equations of particles going in and coming out and used an awful lot of complex mathematics to work out what might have happened in between. He thought it was the ultimate black box.[4] While others around him embraced S-matrix theory, Higgs left it well alone. He had enough faith in his own understanding of quantum field theory to know it wasn't a lost cause. It was his knowledge of the theory that led him to spot the flaw in Wally Gilbert's letter in the summer of 1964 and go on to formulate a theory of mass.

Higgs returned to Edinburgh in August 1966 on a high. His year in Chapel Hill had been pivotal in building his name as a physicist. His theory had barely been noticed until Freeman Dyson took an interest and invited him to speak about it at the Institute for Advanced Study in Princeton. Now, some of the most influential physicists in the world knew about his work. There was good news for Jody, too. Soon after arriving back in Edinburgh, she was offered a job at the university as a phonetics lecturer.

In an imaginary world where science is simple, the future would have gone something like this. Peter Higgs and the five other theorists who worked on the theory of mass would have gone back to their offices, poured themselves a cup of coffee, and wondered what mystery of the universe to solve next. Meanwhile, down the corridor, the experimentalists would have arrived on the scene and unpacked their equipment. In a few short hours, they'd have detected the ethereal presence of the Higgs field, collected some Higgs particles, and promptly declared that they had confirmed the origin of mass. Cue: back-slapping all round.

But the world of science is rarely simple. There was a mountain of work to do before scientists could even hope to test Higgs's theory. For one, the theory he and the others had come up with wasn't clear about which particles it gave mass to. Nor did it say much about the Higgs boson. Particles are easier to find if you know roughly how heavy they should be.[5] The irony of Higgs's theory was that it explained how particles might get their masses but was mute on the mass of the Higgs particle itself. Scientists could go looking for it, but they wouldn't know where to start.

Then there is the Higgs field. Scientists cannot go out and look for the field directly. The Higgs field is effectively hidden in the vacuum of space. What makes it particularly difficult to see is that it doesn't vary from place to place.[6] One thing that makes it easier to study the gravitational field is that it is stronger in some places than others. Climb to the top of Mount Everest and gravity's pull will be measurably weaker than at sea level, because you are farther from the center of the Earth. Theoretically, scientists could induce variations in the Higgs field, but they would need to heat the universe to more than one million billion degrees Celsius. Even if it were possible, it would be highly inadvisable. Changing the Higgs field would alter the sizes of atoms and make everyday matter unstable.[7]

After the 1964 papers appeared, Peter Higgs did the only thing he could. He tried to build a bigger theory than the one he had already. He wrote down calculations and plugged in details of subatomic particles. Sooner or later, he hoped it would all come together in a theory that showed how the Higgs mechanism gave mass to some particles and not others. It was a disheartening process. Months went by without progress. Whatever Higgs tried, he couldn't get it to work.

Things were not going any better in Brussels. Robert Brout and François Englert had the same idea as Higgs, but they weren't making any headway. Try as they might, they could not get their theory to explain the masses of specific particles found in nature. Eventually, they passed the problem on to a young Ph.D. student, but she couldn't do any better. As far as the Europeans were concerned, work on the Higgs mechanism was floundering.

Gerry Guralnik was back in America and had other things to worry about. He feared he might not survive as a physicist. The trashing Heisenberg had dished out in Feldafing had rocked his faith in himself and the theory he had worked out with Dick Hagen and Tom Kibble. The experience had left him "depressed and more than a bit beat up," he said.

Steven Weinberg was thirty-four years old and working at Massachusetts Institute of Technology in 1967. He had taken time off from his regular job as a physics professor at the University of California at Berkeley and moved to Cambridge so his wife could study at Harvard Law School. Weinberg was having a tough time of it. The couple had just moved into their second rented home, and he had taken on responsibility for his young daughter, who was still at nursery school. On top of everything, his work had ground to a halt.

Weinberg was trying to use the Higgs mechanism to explain some subtle differences between protons and neutrons, the ingredients of atomic nuclei. He spent the autumn working with pencil and paper, writing down equations to see where they led, but the answer seemed to be nowhere. He knew it was time to give up when his theory implied that a particle already known to nuclear physicists, the rho meson, was massless. The problem was that physicists knew full well that it wasn't. "It left me very despondent," he said. "Making predictions that are already known to be wrong is no way to get ahead in physics."

Weinberg wrote about his frustration in 1997 in an article for a now-defunct glossy magazine called *George* that was cofounded by John F. Kennedy, Jr.[8] In it, he wrote: "Often, when you're faced with a contradiction like this, it does no good to sit at your desk doing calculations—you just go round and round in circles. What does sometimes help is to let the problem cook on your brain's back burner while you sit on a park bench and watch your daughter play in a sandbox."

A few weeks later, probably in mid-September, Weinberg was driving to the office at MIT in his red Camaro sports car when it dawned on him there was nothing wrong with his theory: he was just interpreting it wrongly. The calculations he had written down had nothing to do with the subtleties of protons and neutrons. Instead, they described beautifully the workings of the fourth force of nature. "I had the right answer to the wrong problem," he says.

The fourth force of nature is probably the least well known of them all. Most people are familiar with gravity and the electromagnetic force, which goes to work inside electronic devices and makes your hair stand up in a storm. The third force is known simply as the "strong force" (it is 137 times stronger than the electromagnetic force), and its job is to hold particles together inside atomic nuclei. The fourth force is even more obscure. Known as the "weak force," it is responsible for certain kinds of radioactive decay. Inside the sun, the weak force converts hydrogen into deuterium (a heavy form of hydrogen), which is the starting point for the fusion reactions that make the sun shine.

One thing that is unusual about the weak force is that it doesn't travel. Whereas the electromagnetic force acts over vast distances, the weak force has a reach of only one hundred millionths of a nanometer. That's about 1 percent of the width of an atomic nucleus. The distance is so small that physicists treat the weak force as though it acts on contact.

Weinberg arrived at his office at MIT and started roughing out the details of the theory. Soon he realized that the massless particle that seemed to wreck his calculations earlier was the photon, the truly massless particle that makes up light and carries the electromagnetic force. This had major implications. It meant that Weinberg's calculations described not only the weak force but the electromagnetic force, too, in one overarching theory. Weinberg had unwittingly unified two of the forces of nature. It was the first time a scientist had achieved the feat since Maxwell unified electricity and magnetism in the nineteenth century.

Weinberg's work described what scientists today call the "electroweak force." His calculations showed that, in the early universe, the electromagnetic and weak forces were intertwined. Then, as the universe expanded and cooled, they were pulled apart into the two separate forces we see today. Weinberg's breakthrough was all the more significant because it was built around the Higgs mechanism. It was the Higgs field that pulled the electromagnetic and weak forces apart.

When Maxwell unified electricity and magnetism, his calculations predicted the existence of electromagnetic waves beyond those that we see as light. Scientists thanked him for it, because it gave them something to look for to prove his theory was correct. As luck would have it, Weinberg's theory made a few predictions of its own. It called for three new kinds of particles, named W and Z bosons. The W (for weak) comes in two forms, positive and negative, while the Z has no electric charge. Its name comes from having zero charge—and also from the fact that Z is the last letter of the alphabet and Weinberg hoped the Z boson was the last of the family of particles that carried the weak force.

The role of the Higgs mechanism was critical for making Weinberg's electroweak theory work. The Higgs field splits the electroweak force in two by making the W and Z particles heavy, while leaving photons massless. Because photons are still weightless, they can carry on transmitting an electromagnetic force over great distances at the speed of light. But thanks to their newfound mass, the W and Z particles can barely move at all, so the weak force can only act over tiny distances. Physicists later realized that quarks and electrons are also caught up in the field, making them massive too.

Weinberg's paper on the electroweak force was published the following month, in November 1967.[9] It became the most cited article in the history of elementary particle physics. The theory said roughly what the masses of the new particles should be, meaning scientists could begin to look for them. If the particles existed in nature, it would be one great shot in the arm for Weinberg's theory and the Higgs mechanism it was founded on. "In a sense I rediscovered the Higgs mechanism," Weinberg recalls. "And now the Higgs particle is the missing element. It's the one thing we don't have."

The year after Weinberg's paper appeared, Abdus Salam, professor of theoretical physics at Imperial College, London, and later director of the International Centre for Theoretical Physics at Trieste in Italy, published essentially the same theory independently. The theories developed by both Weinberg and Salam had similar features to work done in 1961 by Sheldon Glashow, one of Weinberg's former classmates at Bronx High School in New York City. Glashow had roughed out a theory that combined electromagnetism and the weak force. It predicted the existence of a W particle, but it lacked a vital ingredient. The theory did not include the Higgs mechanism, which was still three years away from being published. Without it, the theory didn't work.

Scientists don't always communicate with each other as well as they could. The reasons are very human. They might not meet properly and fall into conversation. Even if they do, they might not mention something that makes an idea light up in another scientist's mind. Part of it is secrecy. Shrewd scientists know it is unwise to talk too freely about their ideas, at least not until they are published. The upshot is that chances to make fundamental advances in science are often missed many times over before they are finally grasped and fulfilled.

Peter Higgs met Sheldon Glashow in 1960 at a summer school for physicists held at Newbattle Abbey College, a stunning sixteenth-century former monastery set on 125 acres of parkland outside Edinburgh. Glashow was twenty-seven years old and had already written down his formula for marrying electromagnetism with the weak force. He was hoping to publish it in the next few months. One night, a group of physicists stayed up late to discuss their latest work. Glashow was among them. If Higgs had been there, too, he would almost certainly have learned about Glashow's idea and could have turned it into a successful theory like Weinberg's. He missed the chance though. That night, Higgs had worked as the wine steward. He had no idea the group was keeping itself well-oiled by stashing the bottles Higgs brought over in the bottom of an old grandfather clock.

In 1979, Glashow gave a high-profile lecture in which he reviewed his own contributions to physics.[10] In it, he wondered why Higgs and the other physicists who worked on the theory of mass never realized that they held the crucial missing piece that was needed to marry electromagnetism and the weak force. He had had plenty of conversations with Higgs and the others. "Did I neglect to tell them about my [unifying] model, or did they simply forget?" he asked his audience. Whatever the reason, the missed opportunity meant physicists had to wait seven years before Weinberg found a use for the Higgs mechanism.

Higgs had only been back in Edinburgh a year when he heard about Weinberg's breakthrough. He read about it with mixed feelings. "I was pleased that someone had found a sensible use for my theory, but it was obviously disappointing. I had been trying to apply it to everything at once and that was a mistake. I was obsessed with the wrong application. Glashow and I simply failed to communicate," Higgs recalls.

What is more surprising is that Glashow didn't beat Weinberg to the answer. The six physicists who wrote down the theory of mass published their papers in the most important physics journals of the time. They were printed only a few years after Glashow's work was. But even if Glashow missed all of their articles, he should have learned about Higgs's theory in 1966. He was in the audience when Higgs spoke about his theory at Harvard, the day after talking at the Institute for Advanced Study in Princeton. Glashow even talked to Higgs afterward and told him he liked the theory. "He hadn't seen that it had anything to do with him," Higgs said. Later Glashow admitted he'd "quite forgotten" the work he had done on the electroweak force.

The missed opportunities didn't end there. After the 1964 papers came out, Gerry Guralnik found himself sheltering from a downpour in a beat-up Ford Anglia with John Charap, a theoretical physicist at Queen Mary, University of London. The two chatted about the theory of mass and hit on the idea of using it to unify electromagnetism and the weak force. For some reason they never looked at it seriously. The idea drifted away with the storm.

On another occasion, Guralnik was having lunch with John Ward, a physicist and collaborator of Abdus Salam's at Imperial College. When Guralnik started chatting about his work there, Ward urged him to stop. He warned Guralnik he shouldn't be so free with his ideas, because someone would steal them before he had finished working on them. "If he had only listened. The two of us had enough information to have had a good chance to solve the unification problem on the spot," Guralnik said. Writing about this some time later, he asked: "How did we miss it? Timidity, slowness and bad luck."[11]

One way or another, all of the physicists involved in the theory of mass in 1964 missed the chance to show how it worked in the real world. Apart from the personal and professional frustration that brought, there was a loss to physics in general. The same problem undoubtedly exists today, with scientists holding different pieces of a greater jigsaw but never quite managing to bring them together in the same place at the same time.

Today, the theory that Weinberg and Salam wrote down all those years ago is a major foundation of the Standard Model that describes the behavior of all of the particles presently known to exist. With the theory, the Standard Model makes sense. It shows how the Higgs mechanism works in nature to give mass to specific particles, quarks and electrons included. Before Weinberg, Higgs's theory was nothing more than an elegant idea. Afterward, it was central to physicists' understanding of matter.

Weinberg's theory didn't set the world of physics alight overnight, though, and for good reason. Scientists feared that his theory suffered from the same problem that nearly sank quantum electrodynamics in the 1940s. Their worry was that, in certain circumstances, Weinberg's theory might churn out results that included infinities. That meant it would predict particles doing things more often than always, which was clearly impossible. Richard Feynman had solved the trouble of infinities in quantum electrodynamics by inventing a technique called renormalization, and Weinberg was sure the same could be done for his theory. The only problem was, he had no idea how to prove it.

The Singel Canal curves like a warm embrace around the ancient town of Utrecht in the Netherlands. Along its bank, three adjacent houses were once home to the university's Theoretical Physics

Institute. It was a curious place to work. At one of the houses you were likely to be greeted by a woman who claimed to be a countess, but surely wasn't. In the summer, chickens from the garden hopped through the windows and walked across the desks. For lunch and coffee breaks, the physicists headed down to the cellar, where the legs of pedestrians were visible through a narrow window that looked out on to the street above. Word had it that in times gone by the building had been used by the town's prostitutes.[12]

Gerardus 't Hooft rented an apartment around the corner from the institute. He had arrived in Utrecht fresh from high school in 1964, the year Higgs and the others published their work on mass. The young Dutchman had set his sights on a career in physics at an early age. As an eight-year-old, his teacher asked what he wanted to be when he grew up. "A man who knows everything," 't Hooft replied. He wanted to say "professor," but he had forgotten that particular word. What he really meant was "scientist": a person who strives to know the fundamental laws of nature.

As a schoolboy 't Hooft demonstrated a rare ability for mathematics, but just as unusual was his way of solving problems. Most gifted students got to the top of the class by learning how to use the theories they had been taught. Gerardus 't Hooft took a different route. He preferred to make his own theories from scratch, a strategy known as working from first principles. It's the equivalent of learning how to build a car, instead of only knowing how to drive it. Working line by line, 't Hooft got to know every nut and bolt of his theories.

The institute had recently taken on a new professor of theoretical physics, Martinus "Tini" Veltman. He became 't Hooft's mentor at the university and steered him through his undergraduate years and on to Ph.D. research. Like 't Hooft, Veltman had a mind of his own, but he was more stubborn and had a decidedly anti-authoritarian streak.[13] Veltman was wary of people who considered themselves experts and always trusted his gut feelings over their advice. When a generation of physicists said quantum field theory was dead, Veltman ignored them and kept working.

Veltman was working on theories like Weinberg's and was determined to prove that the fear of infinities was unfounded. But the work was hard going. Veltman's calculations grew and grew until they included 50,000 separate mathematical terms. They were too unwieldy to write down. Veltman decided that a better way of dealing with the cumbersome equations was to feed them into a computer. Three frantic months later, he had written the necessary computer code and was ready to go.

It was a heroic achievement in an era when computers were programmed by punchcards and took days to churn out their answers. Veltman walked around with hundreds of the cards in his briefcase. The beauty of them was that, even if you dropped them all over the floor and muddled them up, the computer could still read them because they were numbered. Veltman called his program Schoonschip, which means "clean ship" in Dutch. It comes from an old seafarers' phrase that sailors used when they cleaned their vessel from bow to stern before setting off on a voyage. It marked a fresh beginning.

The first theories Veltman tested with Schoonschip were failures. When the program spat out its answers, it was clear that infinities were still an issue. Veltman kept at it, though, tinkering with the theories and putting them back into Schoonschip. Meanwhile, 't Hooft was doing what he did best: working up a theory from first principles. As the theory emerged, he realized that the Higgs mechanism was an integral part of it. He had effectively reinvented Higgs's theory.[14] When he checked his calculations, he saw that the infinities were no longer a problem.

One afternoon in the autumn of 1970, Veltman and 't Hooft took a walk from one of the institute's buildings to another. Veltman was complaining. They needed to draw up just one theory that was renormalizable and described massive particles. "I can do that," said 't Hooft. Veltman stared at him. "What?!" he replied. "I can do that," 't Hooft repeated. The moment is seared on Veltman's brain. He nearly walked into a tree. "Write it down, we will see," Veltman said.

When the two compared notes, it became clear that Veltman had left something out of his theories: the Higgs mechanism. He thought it was a fancy trick and had chosen to ignore it. When he added it and ran the program again, it showed that infinities arose but immediately canceled each other out. The work, completed in 1970, proved not only that Weinberg's theory was on solid ground but that it had the Higgs mechanism to thank for it. The following summer, Veltman organized a conference on particle physics in Amsterdam and gave 't Hooft a ten-minute slot at the end to announce the breakthrough. "We'll astound everyone!" Veltman told 't Hooft. Physicists sat up and took notice. As the Harvard physicist Sidney Coleman put it, "'t Hooft's kiss transformed Weinberg's frog into an enchanted prince."[15]

"You're famous!" Peter Higgs looked up and saw Ken Peach, his friend and a fellow physicist who had walked into the staff club at Edinburgh University where Higgs had just finished lunch. It was the summer of 1972, and Peach had been away at a scientific meeting at the U.S. Fermi National Accelerator Laboratory, or Fermilab, a major physics facility on the outskirts of Chicago. At the meeting, Higgs's name had been plastered over almost every talk that touched on symmetry-breaking and the origin of mass. It was the first time Peter had heard anyone mention things like Higgs fields and Higgs mechanisms so publicly.

Higgs smiled. The news was a welcome boost for his morale. For all the groundbreaking work he had done on the origin of mass, he was finding it increasingly hard to keep up with other scientists in the field. Talking about the situation years later, Higgs said: "Because I'd written an influential paper in getting it all going, people assumed I understood whatever came next. But increasingly I didn't. When it came to what Veltman and 't Hooft did, I wouldn't just struggle, I'd sink."

Higgs decided to move on to other things. He got interested in a theory known as supersymmetry, which paints an extraordinary view of the world. In one fell swoop, it doubles the number of particle species in the universe. Each new particle is a heavy—and as yet unseen—partner of a more well-known particle. There are selectrons paired with electrons and squarks with quarks. Scientists find it compelling because it overcomes some long-standing problems in physics. Some versions of the theory have not one Higgs particle but five, which all play a role in making particles massive.

The increasing complexity of his work was only part of the difficulty Higgs was facing. In the spring of 1972, he and Jody, with whom he'd had another son, broke up (though they never divorced). "I wasn't in any state for getting involved with detailed theoretical work," Higgs recalls. In an interview with the *Sunday Times* in 2008, Higgs blamed his obsession with work for ending his marriage: "We split up because I had put my science career above the family. One time I backed out of a family holiday when we were meant to be going to America. Then I got on a plane and went to a conference. Jody, my wife, just lost touch with what I was doing."

The point in history when Higgs's name was immortalized is contested. Higgs believes it was the 1972 conference at Fermilab. According to Higgs, he alone became best known for the theory because of a chance event. In 1967, Higgs got chatting about his work with the Korean American physicist Ben Lee over a glass of wine and a sandwich at a conference reception in Rochester, New York. It turned out that Lee was the rapporteur at the Fermilab conference in 1972 and had remembered their conversation. When he was writing the program, he used Higgs's name as shorthand for the origin of mass theory and anything remotely like it. From then on the name simply stuck. Dick Hagen believes the name "Higgs boson" was first used at the International Conference on High Energy Physics that was held in Berkeley from August 25 to September 1 in 1966; he even wrote a letter to the organizers after the conference to protest.

If the most important thing in the life of a scientist is to discover unknown truths about nature, then getting credit for those discoveries must be a close second. Careers are built on reputation, and recognition is a necessary part of that. When it comes to a big discovery, due credit can mean promotion, fame, and, sometimes, fortune. With so much at stake, it is unsurprising that tensions run high over who did the crucial piece of work that flicked the switch of any particular discovery. It is often far from clear. Hundreds of scientists might have contributed in one way or another. Work that seems irrelevant one day might turn out to be a missing piece of the jigsaw the next.

In June 1938, George Paget Thomson, who won the Nobel Prize in Physics a year earlier, gave a lecture on his discovery of electron waves, a phenomenon first postulated rather vaguely by Louis de Broglie in Paris in 1924. He explained how discoveries in science rarely came from one person, but were invariably a group effort: "The goddess of learning is fabled to have sprung full-grown from the brain of Zeus, but it is seldom that a scientific conception is born in its final form, or owns a single parent. More often it is the product of a series of minds, each in turn modifying the ideas of those that came before, and providing material for those that come after."

Tensions over due credit run particularly high when the work in question is deemed worthy of a Nobel Prize. Steven Weinberg has said it was a pity the Nobel Committee failed to award Freeman Dyson the prize for his contribution to sorting out quantum electrodynamics in the 1940s. Instead the prize went to those whose work he brought together: Richard Feynman, Julian Schwinger, and Sin-Itiro Tomonaga. And herein lies the contentious issue. The Nobel Committee never awards one prize to more than three people. In this case, it meant that Dyson is one of a sizable and august group of scientists who have cause to feel slighted.

The Nobel Committee is heading for a similar quandary over the theory of mass.[16] The work is widely regarded as of Nobel Prize quality. Weinberg, Salam, and Glashow shared the Nobel in 1979 for their work on the electroweak force. Twenty years later, Tini Veltman and Gerardus 't Hooft won the prize for proving the theory was renormalizable and so not plagued by infinities. In these cases, pinpointing the right scientists was straightforward. The particle at the heart of the theory of mass is known to the physicists and the world's media as the Higgs boson. That makes Peter Higgs's claim to any future Nobel Prize look assured. But there are five other physicists who came up with the theory, two of whom beat Higgs into print. As the scientists involved in the work retired and the discovery of the Higgs boson loomed on the horizon, tensions over the naming of the theory started to emerge.

One physicist tells a story about a lecture he attended on the theory of mass in Brussels some years back. The speaker, Lalit Sehgal from the Institute of Theoretical Physics in Aachen, Germany, switched on his computer, got his slides in order, and started talking about the Higgs mechanism. Not long into the lecture, he noticed that a distinct look of displeasure had come over a man in the front row. Guessing at his fault, he said: "I realize this theory was described by a number of researchers, but in accordance with custom, I refer to it by the name that is shortest." Before Sehgal had a chance to go on, the man in the front row raised his voice. "My name has five letters too!" It was Robert Brout.

The two Brussels researchers, Brout and Englert, are frequently credited for their work on the Higgs mechanism, however, and they bear Peter Higgs no grudge. The third group, Gerry Guralnik, Dick Hagen, and Tom Kibble, get very little recognition, even among physicists who work on the Higgs mechanism. Guralnik and Hagen believe some European physicists are even conspiring to have them written out of history. Guralnik raised his suspicions in an article published in 2009.[17] In it, he wrote: "Initially there seemed to be no problem getting recognition for what we did on a more than equal basis to the Englert and Brout and Higgs papers. This seemed to change around 1999, when our work began to be omitted from the references contained in important talks and papers, even by authors who had previously referenced us."

Robert Brout and François Englert at the Free University in Brussels were the first to publish on what is now widely known as the Higgs mechanism. They were isolated from the international particle physics community and newcomers to the field. Higgs was next into print and the first to draw attention to the existence of a new particle, the Higgs boson, which must exist if the theory is correct. Third to publish were Guralnik, Hagen, and Kibble, with what some regard as the most comprehensive version of the theory.

Higgs is uncomfortable with his name alone being attached to the theory. In conversation, the Higgs particle becomes the "scalar boson," or the "so-called Higgs boson." At one conference, Higgs acknowledged the awkwardness of the situation by beginning his lecture like this: "Contrary to the custom at this conference, I want first of all to disclaim priority for some of the concepts to which my name is commonly attached in the literature."[18] He went on to suggest that the Higgs mechanism should be renamed the "ABEGHHK'tH mechanism" after all the people (Anderson, Brout, Englert, Guralnik, Hagen, Higgs, Kibble and 't Hooft) who discovered it or rediscovered it.

Whatever you call the Higgs mechanism, the work of Glashow, Weinberg, Salam, Veltman, and 't Hooft had shown that it was almost certainly a vital ingredient of a much greater theory that unified electromagnetism with the weak force. Now the task was to see if Weinberg and Salam's theory checked out. As luck would have it, the theory predicted the existence of three kinds of particles that had never been seen before. His calculations showed that the two W particles should weigh around forty times as much as a proton, with the Z particle weighing twice as much again. Knowing their masses made all the difference: it meant physicists had a good idea where to start looking.

5

An Earnest Revenge

The plane touched down with a brief screech of rubber, the undercarriage heaving as it shouldered the full weight of the aircraft. The engines slowed to a steady hum and the plane taxied over to the terminal and came to a halt. Over the intercom the pilot gave the all-clear for passengers to unfasten their seatbelts and welcomed them to a bleak and wintry England.

Inside the terminal building at Heathrow airport, Donald Perkins, a physicist at Oxford University, watched the passengers collect their suitcases and join a queue to have their passports checked. It was December 30, 1972. The inbound flight from Germany was bringing people over for the New Year or home from a Christmas break. Perkins didn't have to look hard to spot the person he was waiting for. Helmut Faissner, a physicist from Aachen University, had seen him already. He was grinning from ear to ear and waving a sheet of photographic film in the air.

When Faissner cleared customs he took Perkins to a table and laid the photograph down so they could both take a look. The black background was covered with white flecks and delicate swirls and rings that looked like gunshot wounds. To the untrained eye it was a meaningless jumble of streaks and blotches, but to Perkins and Faissner the film was enough to make the heart race. Perkins realized its significance immediately and demanded they go directly to the bar and celebrate. Faissner suggested they drive to Christie's, the auctioneers, and sell the photograph to the highest bidder. It was what Faissner called a "Bilderbuch event": a picture-book example of something new to the world of physics.[1]

In 1972, few physicists had heard of the Higgs boson, and those who had knew it was far too early to start hunting for the particle. The reason was simple: physicists had no idea how to find it. To hunt for it with so little information was to go on a wild goose chase. Instead, particle physicists set their sights on finding evidence for the electroweak theory that Steven Weinberg and Abdus Salam had proposed in the 1960s, and that Tini Veltman and Gerardus 't Hooft had put on solid footing in 1971. The importance of the theory was hard to overstate: it was the first to unify two forces of nature since Maxwell had figured out the relationship between electricity and magnetism in the late nineteenth century. Physicists knew that a Nobel Prize was waiting for anyone who found evidence that the electroweak theory was right.

By chance, the theory of mass was intimately wrapped up with the work physicists had in mind. The electroweak theory relied on the Higgs mechanism. It was the Higgs field that gave mass to the new particles it predicted. If the electroweak theory checked out, then the Higgs mechanism, or something like it, was most probably real. It wouldn't count as concrete proof of Higgs's theory, but it would be the first tentative and indirect evidence that the idea was sound.

The electroweak theory made plenty of predictions that physicists could look for in their experiments. There were the two new particles, the W and Z, for a start, which were not quite as mysterious as the Higgs particle. Physicists had some ideas about how to find them, since they knew something about their mass and how they might reveal themselves in experiments. There was also a subtle effect known as a "neutral current" to look for. Everyday electrical currents are nothing more than negatively charged electrons flowing from one place to another. The theory predicted a new kind of current that was created by a Z particle (which is electrically neutral) moving between other particles. If you had the right equipment, physicists believed you could photograph a neutral current in action. It would leave a telltale pattern on photographic film, a pattern that looked just like the one that spiraled across the film Helmut Faissner was holding when he arrived at Heathrow that day.

If the W and Z particles or fleeting neutral currents didn't show up in experiments, the electroweak theory could be written off as junk, and a big question mark would hang over Higgs's idea of how particles gained mass. To find the answer, physicists turned to the workhorses of particle physics: machines that accelerate streams of particles to phenomenal speeds before crashing them into lumps of metal or head-on into other particles coming the other way.

Particle accelerators began life in the late 1920s as ramshackle devices built from spare parts, but they have been transformed over the decades into the largest and most complex machines on the planet. The earliest models produced beams of high-speed particles that were used to break open atomic nuclei. Eventually the machines were powerful enough to create entirely new particles from the energy released in the collisions.

The rise of the machines can be traced back to Ernest Rutherford and other physicists who did similar experiments in the 1900s. Rutherford knew that radioactive materials produced streams of high-speed particles that could be used to study the structure of the atom. One common material he and his contemporaries used was radium. It emits alpha particles, which are made up of two protons and two neutrons, at speeds in excess of 20,000 kilometers per second. It was a beam of alpha particles that Rutherford used in experiments that led to the discovery of the atomic nucleus in 1911.

After his defining experiments in Manchester, Rutherford moved to Cambridge University, where he became an inspirational leader at the prestigious Cavendish Laboratory. In 1927, he gave a presidential address at the Royal Society in which he highlighted the need for physicists to get their hands on ever more powerful beams of particles.[2] "This would open up an extraordinary new field of great value, not only in the constitution and stability of atomic nuclei, but also in many other directions," he said.

Rutherford's words did not fall on deaf ears. At the Cavendish, the Irish physicist Ernest Walton and his collaborator, a Yorkshireman named John Cockroft, had started bolting together equipment that produced beams of particles without using lumps of radioactive material. The device was crude, but it worked. At one end was a glass flask filled with hydrogen. By applying a voltage across the flask, Walton and Cockroft stripped electrons off the hydrogen atoms, leaving them as bare hydrogen nuclei, which are otherwise known as protons.[3] Since protons are positively charged, they could be accelerated by another voltage down an 8-meter-long pipe attached to the flask. If the device worked as planned, the protons would hurtle down the length of the tube and slam into whatever the scientists put in the way.

Walton and Cockroft had at least one eye on safety. Before testing the machine, they huddled inside a small wooden hut they had built in the middle of the laboratory. They had lined the hut with lead shielding. Their device was as successful as Rutherford had thought it would be. In 1932, the physicists pointed their rudimentary particle accelerator at some lithium, the lightest naturally occurring metal on Earth (some stars and planets are believed to contain metallic hydrogen, a substance that has been created in laboratories by subjecting liquid hydrogen to high pressure). The beam of protons slammed into the target and split the lithium atoms in two. Walton and Cockroft shared the Nobel Prize in 1951 for artificially accelerating particles and splitting the atom.

Splitting the atom was a landmark achievement, but physicists needed more powerful accelerators if they wanted to pulverize atoms enough to study their smallest, innermost constituents. Physicists rated their accelerators by the energy of the particles they produced. The units of the trade are called "electronvolts" (eV), where 1 electronvolt is the amount of kinetic (or motion) energy an electron gets when it is accelerated by a 1-volt electric field. An electronvolt is a minuscule amount of energy. It takes roughly 150 trillion electronvolts to lift a dime a millimeter off the ground. To split the atom takes a more modest 100,000 eV. And you can knock an electron off an atom with only 14 eV, only a couple more volts than a car battery can muster. Physicists describe beam energies in multiples of electronvolts, with keV for a thousand, MeV for a million, GeV for a billion, and TeV for a trillion.

One early problem that dogged the first accelerator builders was the difficulty in maintaining the strong electric fields they needed to drive particles to ever higher speeds. In principle, you could drive beams of particles to a higher energy by accelerating them with stronger fields over a greater distance. But in reality the idea failed. The huge electric fields physicists used broke down, causing huge sparks to arc through their equipment.

While Walton and Cockroft persevered with their pipe-based accelerator, an American physicist, Ernest Lawrence at the University of California, Berkeley, pushed ahead with a new design that solved the problem of high electric fields.[4] He got the idea from an article published in a German technical journal by the Norwegian engineer Rolf Wideroe. Lawrence couldn't hope to follow the text, so he worked it out from the diagrams alone. Instead of accelerating particles down a long straight pipe, the method used weaker electric fields to accelerate them around and around in a spiral.

Lawrence built what became known as a "cyclotron accelerator." Inside the machine, which was no bigger than a small plate, particles were propelled around by an alternating electric field that gave them a kick with every revolution. It was like spinning a playground merry-go-round faster and faster by standing on one side and giving it a hearty tug every few seconds. The particles, steered by powerful magnets, looped around inside the machine and spiraled out as they picked up momentum. Before long, Lawrence was accelerating particles to nearly 5 MeV. The best Walton and Cockroft could manage at the time was 800 keV. Lawrence's machine, which he called his "proton merry-go-round," was not only more powerful, but also compact enough to sit on his laboratory bench.

Lawrence built a series of cyclotrons, each one larger and more powerful than the one before. The first was only 5 inches across, but by 1939 his designs had grown to include an ambitious machine that measured 5 feet across. Lawrence used his cyclotrons to bombard elements with protons, creating radioactive versions of them. His work led directly to the use of radioactive substances in medicine. His brother, the physician John Lawrence, used radiophosphorus to treat leukemia. Soon, his colleagues had figured out how to use the beams of neutrons produced in a cyclotron to destroy cancer cells in the body. In 1939, Lawrence won the Nobel Prize for his work on cyclotrons and the discoveries he made using them, including the creation of technetium, the first artificial element that doesn't occur in nature.

As accelerators became more powerful the physicists faced fresh technical challenges. The particles inside the machines were being whipped up to nearly the speed of light. Under these conditions, turning up the energy even more made little difference to the speed of the particles. Instead, and in line with Einstein's relativity theory, the extra energy alters the orbits of the particles and is dealt with by carefully varying the alternating electric field that accelerates them. These more advanced machines became the next generation of accelerators, known as synchrocyclotrons.[5]

Throughout the Cold War, the construction of giant particle accelerators in the United States and the Soviet Union mirrored the technological one-upmanship of the space race. The two nations saw a scientific imperative in pouring money into the machines: after all, it was knowledge of the atom that had enabled the Allies to win World War II. Learning more about the atom, and the energy locked up inside, was a matter of national security, and the countries continued with little regard for the price tag. When one nation built a huge accelerator, the other had to go one better.

The 1950s was the heyday for large accelerators, with more than a dozen in operation or under construction around the world.

Brookhaven National Laboratory at Long Island in New York ran the Cosmotron accelerator at 3 GeV. At Berkeley, the Bevatron accelerator reached a record-breaking energy of 6.2 GeV. In 1957, the Soviets replied with their own accelerator at Dubna, north of Moscow, which produced beams of particles at 10 GeV. The same year they launched Sputnik, the world's first artificial satellite. It had been just thirty years since Walton and Cockroft had built their first accelerator, but the particles produced in the latest machines were already reaching energies 50,000 times greater than those produced in the makeshift equipment the two physicists had pieced together in their lab.

The huge investment in accelerators in the United States and the Soviet Union posed a serious challenge to Europe, where science was languishing in the economic depression that followed the war. The expertise in running the machines and the knowledge of nuclear physics they generated stayed in the countries that owned them. European scientists were losing ground or left to join scientists in the United States.[6]

Concerns over the future of science in Europe prompted leading figures, including two Nobel Prize winners—the French physicist Louis de Broglie and the American Isidor Rabi—to lobby governments to build a huge multinational laboratory. Its role would be to encourage scientific collaboration among nations and keep European scientists at the cutting edge of physics. Several meetings in the early 1950s led to the establishment of a temporary Counseil Européen pour la Recherche Nucléaire (European Council for Nuclear Research), or CERN, to oversee the project. In 1954, a dozen European countries ratified the creation of a European Organization for Nuclear Research, to be based near Geneva, Switzerland. The CERN acronym was later adopted as the name of the new laboratory.

The European laboratory was big on ambition. The first major accelerator at CERN, the proton synchrotron, was 200 meters wide— almost as wide as the National Mall in Washington, D.C. The logbook for the accelerator marks a moment in CERN history: 7:35

P.M. on November 24, 1959, when the machine accelerated protons to a record-breaking energy of 24 GeV. The following morning, John Adams, a future director general at CERN, announced the success clutching an empty bottle of vodka.[7] The bottle had been sent over by scientists at the Dubna accelerator in Russia on the strict proviso that it only be drunk once CERN broke their record. Adams returned the empty bottle, its original contents replaced by a Polaroid showing the 24 GeV pulse of protons on one of the machine's displays.

The CERN laboratory became a major draw for particle physicists in Europe. But the accelerator itself was only part of the project. Before the machine could be used as a scientific instrument, physicists had to build and install detectors to study what happened when the high-energy particles crashed into other materials. The detectors were designed to look for specific new effects, such as neutral currents or signs of the W particle. These were exquisite pieces of engineering in their own right and could take years to plan and build.

Across the Atlantic, there was no let-up in building more powerful accelerators. As CERN was finding its feet, the first multi-million-dollar laboratories, with accelerators measured in miles and kilometers rather than feet and meters, went into operation. The 3-kilometer-long Stanford Linear Accelerator was built at Menlo Park; another major facility, the National Accelerator Laboratory, went into construction on 6,800 acres of prairie land about 40 miles west of Chicago.

At Brookhaven National Laboratory, engineers built the huge Alternating Gradient Synchrotron, for a time the most powerful particle accelerator in the world, with beams running at an energy of 33 GeV. The machine earned scientists three Nobel Prizes. In 1962, Leon Lederman and others used the accelerator to discover a particle called the "muon neutrino." In 1974, physicists at the Alternating Gradient Synchrotron and at the Stanford Linear Accelerator Center jointly discovered a particle they named the "J/psi," which proved the existence of a new kind a quark, known as the "charm quark."

The architect of the Chicago laboratory was Robert Wilson, the former head of experimental nuclear physics on the Manhattan Project and onetime graduate student of Ernest Lawrence. An accomplished sculptor, Wilson regarded himself as a Renaissance man. He saw accelerators as the cathedrals of the modern age and described the role of the new accelerator in Chicago as "primarily spiritual." The National Accelerator Laboratory would be a thing of "great beauty," he said, that would bring "satisfaction to our lives."

Wilson had grown up on his family's cattle ranch in Wyoming and was a competent cowboy.[8] Edwin Goldwasser, his deputy on the Manhattan Project, said Wilson could "deftly lasso any of his three sons when they were young and when needed"—and when machinery broke down on the ranch, Wilson had learned he could save time and money by fixing it himself. The experience left its mark. "I had the confidence that with your own hands you could build large contraptions and make them work," he said.

Wilson became a candidate for running the Chicago laboratory in 1965, when he was working on accelerator designs at Cornell University. At the time, the U.S. Atomic Energy Commission (AEC) was inviting scientists to submit plans for a new machine on the site. One proposal, from Wilson's former laboratory in Berkeley, ended up on Wilson's desk, and he went through it in detail. He panned it, claiming it was overdesigned and, at $340 million, ludicrously expensive. Wilson had a serious point. He feared that the exorbitant costs of accelerators could kill physics off. When it came to big projects, he was frugal. He believed that if something worked the first time, it was overdesigned; it meant it had cost too much and taken too long to build.

Two years later, when the AEC had reviewed more than a hundred accelerator proposals, they called Wilson and asked him to build the laboratory his way. As construction work got under way, Wilson was hauled before the Congressional Joint Committee on Atomic

Energy to face questions over the project. In the hearing, held in 1969, Senator John Pastore (D-RI) asked Wilson to explain how the facility would improve national security. Wilson said it had nothing to do with security, but Pastore persisted. Finally, Wilson explained the value of the machine. "It has only to do with the respect with which we regard one another, the dignity of men, our love of culture," he said.[9] "It has to do with: Are we good painters, good sculptors, great poets? I mean all the things we really venerate in our country and are patriotic about. It has nothing to do directly with defending our country except to make it worth defending."

Pastore's question was unsurprising, considering the lab was being built at the height of the Cold War. Politicians might have hoped the machines would tell them how to make a more devastating kind of bomb, or give them a way of protecting the nation from an outside aggressor, but that wasn't their purpose. The machines uncovered the physics at work in nature; the application of that knowledge might not be clear for decades. The drive to build ever more powerful accelerators undoubtedly benefited the military indirectly, not least by producing scores of physicists and engineers who were highly skilled in building sophisticated electronics.

Under Wilson's management, the accelerator at the National Accelerator Laboratory was completed ahead of schedule, under budget, and with a potential beam energy of 500 GeV, more than twice the energy that had originally been planned for. The corner-cutting wasn't hard to spot, though. The buildings were ramshackle. Tunnels that had been dug out to make a home for the accelerator leaked. Experimental facilities were sparse or simply weren't built. Undeterred, Wilson began sifting through the list of experiments for which physicists had proposed to use the machine. As was the case at CERN, the experiments he chose dictated the kind of detectors that needed to be built next.

Wilson was hailed as a hero in Washington for taking the project by the scruff of the neck and forcing it to completion. But it was a honeymoon that didn't last long. Wilson had $6 million left over after building the accelerator and thought nothing of diverting it into the construction of a small "booster" machine that could double the energy of the accelerator. When Washington found out, they took a dim view of the matter. It didn't help that Wilson started to avoid calls by hiding his phone in a large plant in his office.[10]

At the start of the 1970s, CERN and the National Accelerator Laboratory in Chicago were ready to go head to head for the first time. The American physicists had far more experience, but their lab had been built on a relative shoestring. At CERN, the European scientists were still coming to grips with the new technology and were desperate to reclaim their place among the world's scientific elite. At both labs, the top priority was to discover evidence for the electroweak theory. Though no one could have known it at the time, the race to find it was a warm-up bout for a much greater contest between the two labs: the hunt for the Higgs particle.

Physicists at CERN and the National Accelerator Laboratory set up similar experiments. They used their accelerators to create beams of particles called neutrinos, which travel at close to the speed of light and almost always pass straight through normal matter. On extremely rare occasions, physicists predicted they would glance off other particles in ways that left telltale signs of the electroweak force at work.

At CERN, hopes for a discovery lay with a team led by the French physicist André Lagarrigue and their 5-meter-long detector called "Gargamelle," named after the mother of the giant Gargantua in the sixteenth-century tale by François Rabelais. Gargamelle was a type of detector known as a bubble chamber, an invention responsible for some of the most beautiful images in physics. To prepare the chamber, technicians filled it with 4.5 metric tons of Freon refrigerator fluid; during experiments, a large piston attached to Gargamelle retracted to reduce the pressure inside the detector, leaving the Freon in a precarious state and on the brink of boiling. If a neutrino slammed into an electron inside the detector, the ricocheting electron would hurtle through the Freon, leaving a track of bubbles in its wake. The tracks triggered flashbulbs to fire so that cameras could capture the moving particles on film.

The Gargamelle experiments ran from the autumn of 1972 through the following spring. Now and then, photographs from the detector showed streaks of bubbles that looked tantalizingly like neutral currents, one of the telltale predictions of electroweak theory.[11] Most of the scientists at CERN were unconvinced. They expected neutral currents to be so rare they would hardly ever see them.

That December, Franz-Josef Hasert, a research student at Aachen University in Germany, was checking through some film taken by the camera on the Gargamelle experiment. There was an unusual spiral of bubbles across the photograph that bore all the hallmarks of a neutral current. He mentioned it to his supervisor, who took it to Helmut Faissner, the head of the Gargamelle team at Aachen. Faissner realized it was exactly what they had been waiting for. Soon afterward, he packed the photograph in his briefcase and flew to England to show Donald Perkins, a member of the Gargamelle team at Oxford University.

Physicists on the Gargamelle experiment had taken 1.4 million photographs of particle tracks in their detector. They calculated that in the pile of film that had been accumulating, they could expect to find between five and thirty streaks of bubbles caused by neutral currents. Each photograph had to be examined in detail on a light table. It was tedious but necessary work. "Anyone can imagine what a nightmare it was," Perkins said later. At that point, the Gargamelle team had examined only 100,000 of the images. By the time they had gone through the remaining 1.3 million, they had only two more images that looked like neutral currents.

In the new year, Faissner penned a note to Lagarrigue to describe the image. "The event has excited us a great deal," he wrote. "It is in effect a lovely candidate for an example of a neutral current." As compelling as the photograph was, the physicists at CERN knew they had to record many more like it to be sure of what they were seeing. The scientists began to sense a discovery at their fingertips.

The physicists dug out their old data tapes and rechecked the images to make sure they hadn't missed any similar pictures. Teams across Europe joined in. They prepared huge enlargements so that each member of the group could inspect them. Standing around each promising image, the physicists argued about the meaning of this streak and that swirl. Was it a neutral current, or something else? Any doubts and the image was marked "out" and put to one side.[12]

On July 18, 1973, Lagarrigue was going through the day's mail at CERN when he came across a letter from Carlo Rubbia, who was leading the hunt for neutral currents at the National Accelerator Laboratory in Chicago.[13] Rubbia said he had heard that CERN was about to announce the official discovery of neutral currents. He said his own team had around a hundred clear images of neutral currents and was in the final stage of writing a paper for publication. Rubbia ended the letter with an offer: he suggested both teams acknowledge each other's findings and share the credit for the discovery.

According to an account in the 2008 book *Fermilab: Physics, the Frontier and Megascience*, "Lagarrigue called Rubbia's bluff."[14] He thought the competing teams should make their own announcements. He declined Rubbia's offer and sent a message to say his team planned to make its announcement at CERN in twenty-four hours. The following day, Paul Musset, a physicist on the Gargamelle experiment, gave a seminar at CERN announcing the discovery of neutral currents. Four days later, the group sent their paper on neutral currents to the journal *Physics Letters*. Carlo Rubbia submitted his team's paper to the American journal *Physical Review Letters* two weeks later. The race was on, but it was far from over. The journal editors sent both papers out to referees, and in both cases the refs came back with questions for the authors. The fate of who would get credit for the discovery hung in the balance.

That September, particle physicists gathered in Aix-en-Provence in the south of France for one of the major conferences on the academic calendar. Paul Musset drove down from CERN. Driving into the city, he passed two men lugging heavy suitcases, presumably from the train station to their lodgings. Musset pulled over. "Are you Salam?" he called to one of the men. Abdus Salam, who drew up the electroweak theory independently of Steven Weinberg, answered that he was. "Get in the car. I have news for you," said Musset. "We have found neutral currents."[15]

That month, *Physics Letters* published the Gargamelle team's paper on neutral currents. They had secured the first major discovery for CERN and put the laboratory on the map. Their jubilation did not last long, however. At the National Accelerator Laboratory in Chicago, the members of Rubbia's team had put their own paper on hold while they tinkered with their detector. The adjustments were intended to give better results, but, when the detector was running again, the results looked worse. The patterns that looked like neutral currents before had vanished.

By November, the U.S. team had had a complete change of heart. Their upgraded detector saw no evidence of neutral currents. The physicists quickly drafted a paper that described the U-turn. The conclusion, from experienced scientists at the National Accelerator Laboratory and the universities of Harvard, Pennsylvania, and Wisconsin at Madison, was damning enough to blow a hole in CERN's paper—and its reputation.

A few weeks later, Carlo Rubbia arrived at CERN with a letter for Lagarrigue that broke the bad news. "We have written a paper intended for *Physical Review Letters*, which will soon be submitted. A copy will, of course, be sent to you, but for obvious reasons we wanted to convey our results informally to you before its publication," he wrote. Lagarrigue circulated it around his group. As the news spread, CERN's director general, Willy Jentschke, called an urgent meeting of the members of the Gargamelle group to cross-examine them on their results.[16] If Rubbia's team was right, CERN faced an awful humiliation. The Gargamelle group stood its ground. The physicists had considered everything that might have produced the patterns on the photographs and ruled out all of them except neutral currents. The results were right, they said, whatever the U.S. team had found.

Soon after Rubbia's visit to CERN, his group at the National Accelerator Laboratory was having doubts again. One member of the team, David Cline, had spotted eight events in the detector that looked like solid evidence for neutral currents. Instead of submitting the paper that dashed hopes of finding them, members of the American team revised their original manuscript, which by now had been sitting with the journal publishers for more than four months. The second U-turn highlighted not only the pressure the U.S. team was under but also the immense skill required to fully understand the complex detectors. Wags joked that the scientists were hunting for "alternating neutral currents." Their paper was finally published in April 1974.

The intense effort to discover neutral currents was exhausting for the teams at CERN and the newly renamed Fermi National Accelerator Laboratory, or Fermilab. But the prize was worthy of the effort. It was the first time scientists had found concrete evidence that the electroweak theory was right, and that electromagnetism and the weak force were combined into one in the early universe.

Few scientific discoveries are fully appreciated the moment they are made. At first, they can seem arcane, obscure, and practically insignificant. Then, as the number of scientists working on a problem grows, the relevance of their findings finally emerges. Neutral currents are shaping up to be a prime example. One of the greatest mysteries in the universe is why aging stars don't gently flicker and fade at the end of their lives, but instead explode in a spectacular fashion. These supernovas release as much energy as the sun is expected to emit over its entire lifetime. As death throes go, a dying star is truly dramatic. More than a decade after neutral currents were discovered, scientists began to suspect that they were crucial to the behavior of dying stars. The currents appear to drive supernova explosions, which in turn produce all of the heavy elements needed for life on Earth.

Neutral currents may turn out to play important roles closer to home. Plenty of molecules come in two forms, identical except for the fact that they are mirror images of each other. Scientists refer to them as left- and right-handed forms. One enduring quirk of nature is that the handedness of molecules matters, in some cases profoundly.

The point about the handedness of nature was demonstrated by Richard Feynman in a collection of lectures published in 1965.[17] Feynman described an experiment in which sugar is made from scratch in the laboratory. The sugar molecule is simple, made up of just 12 carbon atoms, 22 hydrogen atoms, and 11 oxygen atoms. What happens if you put the synthetic sugar into some water and add bacteria? Curiously, the bacteria will eat only half of the sugar. Here's why: synthetic sugar contains equal amounts of left-handed and right-handed sugar molecules. The two kinds are chemically identical, but mirror images of one another. In nature, for reasons unknown, sugar molecules only come in the right-handed variety, the only kind that bacteria will eat. If the bacteria encounter left-handed sugar, they leave it alone. They don't know how to use it.

Handedness is literally built into our genes. The double helix of our DNA is right-handed. The amino acids that make proteins in living organisms are left-handed. The origin of this handedness in nature is one of the most baffling puzzles in biology.

In 1984, Stephen Mason, a chemist at King's College London, found what might be the answer. Particles such as electrons and quarks have a property called "spin" that can be left- or right-handed. The force transmitted by Z particles only affects particles with a left-handed spin. Mason's calculations showed that, when this was taken into account, left-handed amino acids and right-handed sugars were more stable than their mirror-image counterparts. It could be that before life even emerged, neutral currents made some molecules more stable than others, and they gradually became the dominant forms in the universe.

The handedness of nature is more than just a scientific curiosity. Understanding why left- and right-handed molecules behave differently is a pressing medical question. In the 1960s, many women took the drug thalidomide during the early stages of pregnancy to help cope with morning sickness. More than 10,000 children are estimated to have been born with major defects as a direct result. Many were missing arms and legs, but the drug also caused eye and ear malformations and harmed the development of the babies' hearts, kidneys, digestive systems, genitals, and nervous systems. A single dose of the drug was enough to do such damage. The drug was administered as a mixture of left- and right-handed molecules, but scientists now know that only one form helped with morning sickness, while the other caused birth defects, probably by interfering with the activity of genes in the developing baby. The drug would have been safe if it had contained only one type of molecule.

The role of neutral currents in life on Earth intrigued Abdus Salam long after he proposed the unification of electromagnetism and the weak force. In a 1988 lecture in memory of Paul Dirac, he said: "There is a growing confidence today that the electroweak force is the true 'force of life' and that the Lord created the Z particle to provide handedness for the molecules of life."

CERN had notched up its first major success with the discovery of neutral currents. These fleeting hints of the electroweak theory at work convinced scientists that the W and Z particles were there for the taking. Finding them was the next obvious goal for particle physicists and would constitute a major step in putting the electroweak theory on rock-solid ground. The more evidence there was for the theory, the more physicists began to believe in the Higgs mechanism that seemed to underpin it.

One clear problem stood in the way. The W and Z particles are so unstable that they don't hang around for long in the universe today. If they were going to observe them, physicists had to make them, and that was not going to be easy. No accelerator in operation or under construction in the 1970s could generate enough energy to create the particles. The physicists had two options: either wait until they had the time and money to build a new generation of larger, more powerful accelerators, or come up with a quick and dirty alternative.

The first accelerators used high-energy beams of particles to break matter down by brute force. The smaller the fragments you wanted to study, the more energy you needed. To knock an electron off an atom took relatively little energy because electrons were bound to the atom by everyday electric forces. The nucleus was a tougher nut to crack, because the particles inside, the protons and neutrons, were bound together by the (much stronger) strong force.

The next generation of accelerators had to reach higher energies for more subtle reasons, too. Most of the particles that physicists were interested in didn't exist freely in nature, and so had to be created instead. Einstein showed how this was possible with the equation $E = mc^2$. In the new machines, particles were made directly from the energy released when high-energy beams slammed into their targets. According to the theorists, a total collision energy of at least 92 GeV was needed to make Z particles, and around 160 GeV was needed to make W particles (since their mass is equivalent to 80 GeV each and they come in pairs).

In the late 1960s, Gersh Budker, a gifted and entrepreneurial physicist at the Institute of Nuclear Physics in Novosibirsk in Siberia, mooted a radical idea for vastly ramping up the energy of accelerators. Budker and his colleagues ran their facility on a capitalist model, selling off accelerators and associated parts to whoever would buy them to keep themselves in business. Instead of using accelerators to slam particles into fixed targets, Budker conceived of an accelerator that would steer beams of protons into the path of oncoming antiprotons, their antimatter equivalents. The energy released on impact would be huge.

The idea of forcing particles into head-on collisions was not new. Several accelerator teams had already investigated the possibility and proved it could work. The advantage is clear if you imagine the particles as cars. If a driver shunts into a parked car, plenty of energy is used up pushing the stationary car forward. It is the same in particle accelerators. When high-energy particles slam into a fixed target, a good deal of energy is wasted by pushing the particles in the target backward. In a head-on collision, things are very different. When two high-speed particles collide square on, they effectively stop dead, releasing almost all of their energy in a form that can be turned into new particles.

The genius of Budker's proposal lay in the use of antimatter. An antimatter particle has exactly the same mass but an equal and opposite electric charge to its ordinary matter counterpart. That meant you could put protons and antiprotons in the same accelerator and get both hurtling around in opposite directions without having to make finance-busting modifications to the machine. All you needed to do then was to cross the beams inside a detector and take snapshots of the debris thrown out of the violent collisions.

There was one hurdle that threatened to scupper Budker's idea. Antimatter is difficult stuff to handle. Scientists had no idea if they could turn antimatter into the kind of clean, high-energy beams that were used in accelerators. Budker in Novosibirsk and CERN's leading accelerator physicist, Simon van der Meer, worked on the problem and reached what looked like a promising solution. They showed it should be possible to make "cooled" beams of antiprotons that were sufficiently intense and stable for the collider experiments. Beams of normal protons were made by accelerating hydrogen nuclei straight from a canister of the gas, and it was easy to get them all moving in the same way. Antiprotons were made by smashing protons into a metal target and collecting the occasional antimatter particle that flew off. Antiprotons made this way were a mess of different energies. Cooling them, the scientists concluded, should, in principle, turn them into fine, clean beams.

In the summer of 1976, Fermilab physicists had five proposals on the table to convert their accelerator into a proton-antiproton collider. Two were from teams led by Carlo Rubbia. After discussing the plans in detail, the overseeing committee rejected all of them, on the grounds that it was too early to decide the future path the lab should take. Instead they ordered that more studies be done. On the back of the recommendation, Wilson requested $490,000 to build a small-scale machine to test whether it was possible to cool beams of antimatter.

Rubbia was unimpressed. So unimpressed, in fact, that he crossed the Atlantic and put his plan to CERN. The European laboratory had just switched on its latest particle accelerator, the Super Proton Synchrotron (SPS), which propelled particles around an underground ring that was 7 kilometers in circumference and crashed them into a hard metal target. At CERN, Rubbia found an appetite for risk. The research director-general of the laboratory, Léon Van Hove, feared Fermilab was edging ahead—the Chicago accelerator was running at an energy of 500 GeV, some 100 GeV more than CERN's machine. Van Hove called for an all-out effort to turn the lab's machine into a proton-antiproton collider. It was the lab's best chance to find the W and Z particles first, he believed. In a memo to staff, he wrote that CERN was in danger of "repeating at improved level the experiments already done or in progress at Fermilab."[18] They had to take risks, to set their sights on "more than bread and butter physics."

While Van Hove was research director-general, the engineer John Adams was jointly in charge at CERN as executive director-general. The two argued furiously about Rubbia's plan. Adams feared that the European countries that funded CERN would react badly to the proposal. They had already paid for a new particle accelerator, which had only just been switched on, and now CERN was going to ask them for a radical and risky upgrade? Van Hove was adamant: Rubbia's plan was the best option they had to make new discoveries first, and CERN should pursue it with urgency. In one meeting, discussions over the plan became so heated that Van Hove reminded Adams who ran the scientific program. He threatened to resign if Adams didn't back the plan.

Van Hove got his way. At CERN, engineers and scientists got together to see if such a massive overhaul of their accelerator was possible. The job was enormous. They needed to prove that antimatter could be controlled well enough to be of use in the accelerator. If they could prove that, they needed to build new facilities to make antimatter and store the exotic particles. Within a year, engineers had good news to report. They had extended the lifetime of anti-matter in the lab from a couple of microseconds to thirty-two hours. They had then cooled the antimatter into clean, intense beams. Van Hove's gamble looked to be paying off.

On June 8, 1978, Adams marked the achievement in extraordinary fashion. He jotted down a poem about Rubbia and van der Meer's efforts and sent it out as a memo.[19] The poem—too offensive to reprint here—suggested that Rubbia had exploited van der Meer's brilliance to further his own career.

One month later, staff at CERN crammed into an auditorium to hear Adams speak. He praised the successful antimatter experiments and gave particular credit to Simon van der Meer, who had masterminded the work. "It enables a new step to be taken in accelerator design which previously was thought impracticable," Adams said. By ramming protons into antiprotons, the collider would generate a maximum energy of 540 GeV. Adams quipped that the lab might be broke, but it was rich on ideas. He ended his pep talk by saying, "I should add that the entrepreneur of the . . . facility is Carlo Rubbia, another CERN staff member and well-known transatlantic commuter."

The thinly veiled dig at Rubbia went down badly among some of the staff at CERN. The following month, Adams was forced to issue a public apology for his sarcastic remarks.[20] In a memo to staff, he wrote:

> I described his role as that of an entrepreneur, which in current English industrial jargon means the person who first sees the benefits and possibilities of a scheme and then drives it through to a successful conclusion. Unfortunately, it seems that "entrepreneur" has other less favourable meanings and my remark has been interpreted by some people at CERN as a derogatory one. Since this was the

opposite of my intention in making the remark I hasten to point out that C. Rubbia was the principal initiator of the ppbar facility, that his work in developing the project with S. van der Meer was crucial to its final adoption by CERN, and that his enthusiasm for the project and the energy he is devoting to its realisation are both essential elements for its success.

Work to convert the CERN accelerator was soon under way. A major addition to the sprawling laboratory was a system for making and storing antimatter. The antiprotons were created by slamming a beam of protons into a metal target. For every million protons that hit it, one antiproton was made. These were siphoned off and cooled into beams, ready to be fired into the machine.

To watch the particle collisions, CERN dug two huge caverns around the accelerator to make room for detectors. The first detector was enormous. Designed by Carlo Rubbia, it was highly complex and weighed more than 2,000 metric tons. The second, from a team led by the French physicist Pierre Darriulat, was smaller, simpler, and cheaper. Rubbia's detector was the Goliath to Darriulat's David, but CERN called them UA1 and UA2, after the "underground areas" they were installed in.

The construction work was in progress when Peter Higgs arrived at a conference in Geneva in 1979 and took the opportunity to visit CERN. He was taken on a tour of the site and shown the gaping hole in the ground where the antimatter facility was going to be built. At the time, his work on supersymmetry in Edinburgh wasn't going so well. It seemed that the only people who were doing anything worthwhile were the ones who had just received their Ph.D.s. "They could do things in days that took me weeks," Higgs said.

Einstein once said, presumably only half-joking, that "a person who has not made his great contribution to science by the age of thirty will never do so." There's little evidence to back up Einstein's comment, but younger minds are surely more agile. It was a problem Higgs was feeling around the time he visited CERN. "I was doing really stupid things a lot of the time, so after a while I gave up. I was sad about it. I was not competing and I had to recognize that."

On the outskirts of Chicago, Fermilab was in crisis. The tensions between Robert Wilson and the bureaucrats in Washington had spiraled out of control. Wilson was set on building a booster ring for the accelerator at Fermilab that used superconducting magnets to reach a record-breaking energy of 1,000 GeV. Wilson took it for granted that Washington would fund his scheme, but instead he was called in to pitch the plan. He objected, arguing that his job was to run accelerators, not go cap in hand to bureaucrats. The conflict reached an impasse, leaving the project dead in the water. On February 9, 1978, Wilson quit. In his resignation letter, he bemoaned the poor funding that had left Fermilab running at half capacity. The lab's ability to compete with CERN, which enjoyed twice as much funding, was being seriously undermined. "Our scheme to leap-frog their financial advantage by increasing the Fermilab proton energy to 1,000 GeV through the application of superconductivity has been confounded by indecisive and subminimal support," he wrote.

In his final months in the job, Wilson began working on a parting gift to the lab: a 10-meter-high steel hyperbolic obelisk. In planning the sculpture's construction, he discovered, to his horror, that the bill from professional welders alone would be $20,000. Wilson's reaction was true to form: he would do the welding himself.[21] There was one glitch. The laboratory's on-site welders said he couldn't do it. "Why can't I? I'm the director," Wilson said. "I can do anything I please." The welders explained that if he did they would walk out. The welders' shop was a union enclave, and Wilson wasn't in the union. Wilson duly joined the union, signed up as an apprentice welder, and worked on the sculpture during every spare moment he could muster.

Wilson named the sculpture *Acqua alle Funi*. The phrase came from a story Wilson recounted from sixteenth-century Rome. In complete silence, a crowd of dignitaries watched as nearly a thousand men with horses pulled on ropes to raise an Egyptian obelisk into an upright position. Halfway through the proceedings, the heat from the midday sun warmed the ropes, making them stretch and slip. The obelisk began to sag dangerously, and a Genoese sailor in the crowd broke the silence, shouting "Acqua alle funi!" meaning "Water to the ropes." The men struggling with the obelisk were ordered to do this, and the ropes tightened and regained their grip.

The evening Wilson's obelisk was put into place and christened, he decided to watch the ceremony from a small boat on the pond outside Fermilab's main building. Wilson climbed in with a bottle of champagne and was joined by Norman Ramsey, the chairman of the Universities Research Association and the person to whom Wilson had handed his resignation letter. When there was a hitch in the proceedings, Wilson couldn't help but yell ashore: "Water to the ropes!"

Fermilab desperately needed a fresh and visionary leader. The man for the job was Leon Lederman at Columbia University. Lederman was highly regarded as a deeply passionate physicist and had made his name by discovering new particles. He was also a born joker with a knack for boosting the morale of his co-workers when times were tough. During a stint at Brookhaven National Laboratory, Lederman was once overseeing an experiment that required a large, thick metallic shield. Lederman had somehow secured an old navy cannon for the job, but, on delivery, he realized there was a problem with it. Inside the steel tube was a rifling groove that might interfere with the experiment. Lederman commandeered a lightly built student and sent him inside the cannon with a supply of steel wool to stuff into the offending groove. The student lasted for half an hour before crawling out. "Enough," he said. "I quit." Lederman implored the young physicist: "But you can't quit. Where will I get another student of your calibre?"

Lederman is a romantic when it comes to physics. He has written about the day-to-day job as being "filled with anxiety, pain, hardship, tension, attacks of hopelessness, depression and discouragement." All of this was true, Lederman said, but every minute of it was worth it for those rare moments of discovery. The best discoveries always seemed to come in the early hours, he said, when most people were still asleep. "You look and look, and suddenly you see some numbers that aren't like the rest—a spike in the data. You apply some statistical tests and look for errors, but, no matter what you do, the spike's still there. It's real. You've found something. There's just no feeling like it in the world."

Lederman arrived to find Fermilab in disarray. The staff wanted to know where their future lay. Would the lab pursue Wilson's goal of building the superconductor booster ring, and so create the world's first 1,000-GeV particle collider, or was it better to go head-to-head with CERN and convert the main ring of their existing accelerator into a proton-antiproton collider? To settle the matter, Lederman arranged what he called an Armistice Day shoot-out. Anyone with a view on the future of the lab was invited to present their case. Lederman brought in three wise men—Boyce McDaniel from Cornell University, Matthew Sands from the University of California at Santa Cruz, and Burton Richter from the Stanford Linear Accelerator Center—to act as judges. Their role was "to embarrass the advocates as much as possible with penetrating, incisive questions."

The meeting began at 9 A.M. on November 11, 1978, and ran through until 3 A.M. the next morning. Exhausted from yelling at each other, the physicists called the meeting to an end and waited anxiously for Lederman's decision. The following morning, over salmon bagels and coffee, Lederman talked through the arguments with the three wise men. Converting Fermilab's main accelerator ring into a collider would give them a shot at beating CERN to the W and Z particles, but Lederman was unconvinced by the plan. Instead, he chose to throw everything into building the superconductor booster ring Wilson had in mind, and turn that into an even more powerful collider. The machine, named the Tevatron, became one of the most impressive workhorses of American experimental particle physics and one of the few colliders in the world that had a shot of finding the Higgs boson.

Lederman's decision did not leave CERN in a one-horse race for the W and Z particles. At Brookhaven National Laboratory, plans were afoot to build a new collider specifically to hunt for the predicted particles. The machine, called "Isabelle," relied on experimental superconducting magnets and was due to be switched on in the early 1980s, around the same time as CERN's upgraded machine. The work at CERN progressed well, though. The antimatter facility went into operation. The detectors were in place and being tested out. The beam-cooling system was up and running. In July 1981, after a heroic three-year effort, the converted accelerator slammed its first antiprotons and protons together. Within hours, the detectors were taking their first snapshots of particles meeting antiparticles and vanishing in a hefty puff of energy.

At Brookhaven, things were going far less smoothly. The Isabelle collider had run into serious difficulties. The superconducting magnets were harder to design than expected. When physicists at CERN began using their new machine to look for new particles, the physicists at Brookhaven were still working with scale models. Meaningful tests of the collider were years away.

With Fermilab and now Brookhaven out of the running, the race to discover the W and Z particles came down to the two teams at CERN. Although both Carlo Rubbia and Pierre Darriulat were at the same laboratory, they had separate groups of physicists and used different detectors. The nature of the game at CERN was that groups working on different detectors kept their data very close to their chests. This ensured that if one team made a discovery, the other was there to verify or contradict it independently. It also ensured that the teams developed loyalty and competed against one another.

In August 1982, senior staff members at CERN were told to prepare for an important visitor. The name was kept secret, but the announcement was worded strongly enough to prompt the physicists to carry out a thorough inspection of the machine in case anyone had hidden a bomb there. The VIP was Margaret Thatcher, then prime minister of Britain. She had arranged a private visit to the lab in a wind-down period following the end of hostilities in the Falklands War.

Thatcher had trained as a chemist at Oxford University and insisted she be treated as a fellow scientist. She was told about the hunt for the W and Z particles and how they would bolster evidence for the electroweak theory that unified electromagnetism and the weak force. During the day's tour, a scientist on Carlo Rubbia's team, Alan Astbury, gave a short presentation and said: "If we are lucky, and there is a Father Christmas, we will see the W by the end of the year." On hearing this, Thatcher pointed a finger at Astbury. "Right," she said. "I will phone you in January to see whether you have found it." She did not elaborate on what she would do if the physicists failed. Before leaving, Thatcher made the director-general of CERN, Herwig Schopper, promise to contact her personally as soon as the particle was discovered.[22] "She did not want to learn about it from the press," Schopper recalled.

By November, CERN's particle collider was operating at a sufficiently high energy to create W particles if they truly existed. From a billion collisions, Rubbia and Darriulat took a million to analyze in detail. The W particle itself is so unstable it vanishes almost as soon as it is created, so the CERN teams looked for the by-products of its demise: an electron (or its antimatter equivalent, a positron) and a neutrino. Because neutrinos fly through detectors without a trace, scientists deduce their presence by looking for lost energy in the aftermath of the collisions. When neutrinos fly out, they take their energy with them.

Before the year was out, the CERN teams had spotted a handful of collisions that looked as though they had caused a W particle to pop into existence. It could take months to run all the necessary checks. Aware that Christmas was looming fast, Herwig Schopper sent a message to Margaret Thatcher at 10 Downing Street. He reminded the prime minister of his promise to get in touch as soon as the W and Z were found and wrote: "I should have liked to combine seasonal greetings with the report that such a discovery had indeed been made, but, in the absence of incontrovertible evidence, I am nevertheless pleased to inform you, in strict confidence, that results recently obtained point to the imminence of such a discovery." He signed off by assuring Thatcher that he would be in touch directly as soon as "final and irrefutable evidence is available."

The following January, in 1983, Darriulat and Rubbia presented their latest results in public for the first time at a physics conference in Rome. Rubbia described five collisions that looked like they made W particles, but he stressed that the results were "very preliminary." One slide in Rubbia's talk was captioned "Si sono rose, fioriranno," or, "If they are roses, they will blossom."[23] Darriulat spoke next. He was similarly cautious and highlighted four collisions that seemed to show W particles in action. As it happened, Leon Lederman was there to give the closing speech at the meeting. The discovery of the W and Z looked imminent, he said, but he urged his fellow physicists to maintain their skepticism until the results were more convincing.[24]

One week later, Rubbia and Luigi di Lella, a physicist from Darriulat's team, went over their results in front of an audience of CERN scientists in the lab's auditorium. Rubbia spoke on the first day and emphasized that it was crucial for the competing teams to "march together" toward their prized discovery. Di Lella spoke the next day and said his team needed to firm up their results before they could confidently claim a discovery. Directly after di Lella's talk, Rubbia called his team together. He had decided it was time to publish their results. He pulled out a draft paper summarizing their findings so far and said that anyone who wanted to make changes had to do so before the day was out. The manuscript didn't claim the W had been discovered, but laid out the evidence so far. At the end of the day, the manuscript was passed to the editor of *Physics Letters* at CERN.

The following day was a Saturday, but the nature of physics meant CERN was still buzzing with life. That morning, Carlo Rubbia was making his way to the canteen when he bumped into di Lella, and the two began chatting. According to a 2001 essay by John Krige, a science historian, Rubbia said both teams should remember the risk they ran if they published too early.[25] "If you want to publish it's your business, but if those events aren't Ws your career is going to be ruined," he warned. "No, my God no," said di Lella. "We're not crazy, we're going to think it out." Rubbia's paper was already being reviewed.

On Monday morning, Rubbia got word that the paper was approved for publication. Immediately, a member of his team flew to Amsterdam with the article to deliver it by hand to the publisher. Herwig Schopper, who was on business in Japan, received a Telex saying the paper had been accepted. He quickly sent a fax to Margaret Thatcher to say the W particle had been found and flew back to CERN. The next day, Schopper called a press conference, with Rubbia and van der Meer on one side and Darriulat on the other. A press release issued the same day announced that Rubbia's paper confirmed the discovery of the W particle. The news appeared the next day, making the front page of the *New York Times*.

Darriulat's paper was published three weeks after Rubbia's. Writing about the race in 2004, Darriulat bore no grudge and defended his team's caution. "I am proud we resisted the pressure that was exerted on us to publish faster than we thought we had to. It would have been stupid and childish to give in, and would not have shown much respect for science. . . . The issue at stake was not to bet on the truth, but to behave as if we were the only experiment."

In the summer of 1983, six months after the discovery of the W particle was made public, CERN announced that Rubbia's team had also discovered the Z particle. The telltale signature showed the particle collapsing into an electron and a positron, which hurtled away from each other at high speed. Again Darriulat's team was there to confirm the discovery soon afterward. The finding completed an impressive hat-trick for CERN and dispelled any remaining doubts about the laboratory being a major contender on the world stage.

The discovery of the W and Z particles was the evidence physicists needed to confirm the electroweak theory. It was a healthy shot in the arm for Higgs's theory, too. The electroweak theory only makes sense if the Higgs mechanism, or something very much like it, is also real. With neutral currents and the W and Z particles in the bag, it was time for particle physicists to get the Higgs squarely in their crosshairs.

In the lull between finding neutral currents and recording the first tracks of W and Z particles, theorists at CERN worked out the first detailed description of what the Higgs particle should look like if it turned up in a particle collider.[26] In forty-eight pages, they laid out the physicists' equivalent of a police photofit mugshot. Instead of describing facial features, the paper explained how the Higgs particle would be made in collisions, what kinds of particles it would collapse into, and the chances of seeing it in different machines. Finally, it calculated roughly how long the particle would survive once created, a figure that could be anything between 600 microseconds and one hundredth of a billionth of a billionth of a second.

The paper, written by John Ellis, Mary Gaillard, and Dimitri Nanopoulos at CERN, opened on a cautionary note. "The situation with regard to Higgs bosons is unsatisfactory. First it should be stressed that they may well not exist," the scientists wrote. They ended with an apology to the experimentalists working on the collider for having no idea what the mass of the Higgs boson was. The paper went on to anticipate the technical difficulties in finding the Higgs particle at all and concluded: "For these reasons we do not want to encourage big experimental searches for the Higgs boson, but we do feel that people performing experiments vulnerable to the Higgs boson should know how it may turn up." The paper could not have been more cautious, but it gave scientists their first clues about how to hunt for the Higgs particle.

The run of discoveries at CERN had a profound impact on particle physics in America. The announcement that the Z particle had been found at the European lab came days before a major week-long meeting of the U.S. High-Energy Physics Advisory Panel (HEPAP), which planned research facilities. The news focused minds. American physics was losing ground and needed a dramatic shake-up. George Keyworth, a physicist at Los Alamos who became Ronald Reagan's science adviser following his election in 1980, said America's world leadership in high-energy physics had declined. "In the years American physicists squandered on a pork barrel squabble, the Europeans moved boldly ahead," he said.[27]

The mood was captured in a *New York Times* editorial published in June 1983 as the physicists arrived at Woods Hole, Massachusetts, for the second day of their meeting. The headline read: "Europe 3, U.S. Not Even Z-Zero," and the piece praised CERN for confirming beyond doubt the theory that unified electromagnetism with the weak force.[28] "The bad news," the *Times* said, "is that Europeans have taken the lead in the race to discover the ultimate building blocks of matter."

The editorial acknowledged that "national comparisons have little meaning in physics," since Americans had worked at CERN and other European labs, but stressed the importance of competition in advancing knowledge, asserting that American accelerators should be built to win or not built at all. The three points Europe had scored were CERN's discoveries of the W and Z particles, together with the finding at a German accelerator of gluons, the aptly named particles that stick quarks together inside of protons and neutrons. The editorial said the scoreline called for "earnest revenge" against Europe.

A month later, HEPAP published its recommendations. Isabelle, the $500 million Brookhaven accelerator, was axed. American physicists needed a machine that would get pulses racing, a machine so powerful it would blow CERN out of the water and dominate the field for decades. The machine was to be called the Superconducting Supercollider. It would slam beams of particles together with a staggering energy of 40 TeV, some four hundred times more than was initially planned for CERN's next collider, which was still on the drawing board. The awesome new machine was built to find the Higgs particle or whatever took its place.

In an article about the Superconducting Supercollider for the U.S. magazine *Physics Today*, Leon Lederman and Sheldon Glashow warned that if the United States failed to build the machine, "the loss will not only be to our science but also to the broader issue of national pride and technological self-confidence. When we were children America did most things best. So it should again."

6

Reagan's Renegade

Alvin Trivelpiece, a onetime poker partner of Leon Lederman's, stood outside the Cabinet room in the West Wing of the White House with an oversized easel bearing big, bright pictures.[1] It was a gloomy January afternoon in 1987, and word had gotten around that President Reagan's eyesight wasn't what it used to be.

As director of the Office of Energy Research, Trivelpiece had been granted fifteen minutes to win the president's support for the largest and most costly atom smasher ever conceived. A green light, the advocates said, would guarantee American dominance at the forefront of high-energy physics for decades to come. Without his backing for the project, the nation's historic leadership in unraveling the nature of matter was sure to fade as other countries pushed on.

The Superconducting Supercollider sounded like the kind of diabolical weapon a comic-book super-villain might build in his (or her) lair to hold the world to ransom. In practice, it was the world's first particle accelerator to be designed specifically to look for the Higgs boson.[2] By selling the machine as a Higgs hunter, physicists gave tacit admission that the elusive particle was the most important prize in high-energy physics.

The supercollider wasn't the only machine that had a chance of discovering the Higgs boson. At Fermilab, Lederman's Tevatron had been colliding protons and antiprotons since 1985, though at energy levels too low to prove the existence of the Higgs particle. At CERN, engineers were building a new machine, the Large Electron Positron (LEP) collider, and were expecting to switch it on within two years. Both machines would need major upgrades before the scientists would have a realistic shot at discovering the Higgs boson, but at least they were up and running. In the particle accelerator business, that is no trivial achievement.

The supercollider was supposed to have been a dream machine for American particle physicists, but it is remembered as a nightmare. The project stands out as one of the darkest episodes in the history of the hunt for the Higgs particle, and as a warning of how, when big ideas fail, they can do so spectacularly.

Trivelpiece had set off early for the meeting, but along the way fell into conversation with William Martin, the deputy secretary of energy. Martin wasted no time reminding him that a lot of time and money had been spent arranging the meeting. He went on to add that all of Alvin's friends and colleagues were relying on him to win the president over. "Now don't be nervous," Martin said, as he turned to leave. Right up to that moment, Trivelpiece hadn't been nervous at all.

At the White House, Trivelpiece set up his easel and put briefing notes on the table as the room began to fill up. When President Reagan arrived, Trivelpiece took the opportunity to have his photo taken and then launched into his presentation. The president and the cabinet members looked on. Trivelpiece knew enough about Washington's corridors of power to spare the assembled politicians a dry lecture on fundamental particle physics. To gain their approval to build the Superconducting Supercollider, he took a gamble and turned to something he hoped they would be more familiar with: guns and bales of hay.

The talk began conventionally enough. There were big pictures of tiny fireballs created as subatomic particles crashed together at immense speeds. There were diagrams showing the known atomic constituents. With ever more powerful accelerators, Trivelpiece explained, scientists had peered deeper into the atom. There, they had found a smear of electrons, an enormous amount of empty space, and a hard central nucleus, itself a bundle of protons and neutrons, each built up from quarks.

For all they had discovered, though, there was still a morass of unknowns, Trivelpiece said. One gaping hole in their understanding was that no one knew for sure what made the most fundamental particles weigh anything. The origin of their mass was believed to be linked to the Higgs boson, but the theory had never really been tested.

Trivelpiece pulled the last sheet from one of his easels to reveal a blank page. "From time to time, you will have read classified documents that say 'this page left intentionally blank,'" he told the room. "I want you to try and imagine this isn't a blank page, but a bale of hay. Somewhere inside it, there is a bunch of billiard balls." The president and his advisers did their best to imagine.

Trivelpiece pressed on. "Without breaking the hay bale apart," he said, "how do you work out where the billiard balls are? How do you work out how big they all are?" The room was a sea of blank faces.

Trivelpiece explained that you could take aim with a BB gun and fire pellets at the hay bale, but the shots might not go far enough in. "But what," he said, "if you loaded up a rifle, stood back, and shot high-velocity bullets into the bale every half-inch or so?" Some of the bullets would shoot right through, while others would deflect off the balls. By looking at where the bullets went in and where they came out, it would be easy to work out the size of the balls and where they were hidden. This, he said, was effectively what physicists were doing with particle accelerators.

He showed a picture of Fermilab, home to Lederman's Tevatron accelerator, pointing out it was in Illinois, not 50 miles west of Chicago, as most people thought of it, but also 50 miles east of Dixon, the president's place of birth. The president smiled while others in the room groaned.

The Superconducting Supercollider would be the most powerful particle physics gun in the world, with which they could look deeper inside the atom than any other nation could hope to for decades. The machine was expressly designed to find the Higgs boson, but would almost certainly discover new phenomena no one had predicted. At the time, the estimated cost of building the supercollider was $4.4 billion. By Trivelpiece's reckoning, that meant each minute of his pitch had to earn more than $300 million.

The Cabinet fell into discussion about the machine, some arguing against it, others for. American particle physics was in decline and the machine could rescue it, for sure. But the nation was saddled with a vast budget deficit and the sums at stake were enormous.

Sitting on President Reagan's right side was James Miller, his budget director. He turned to his boss and said: "If you approve this, you are only going to make a bunch of physicists happy." President Reagan turned back and replied that he probably should make them happy, since he made his own physics professor deeply unhappy when he was in school.

The Cabinet members took turns to give their views until finally all eyes settled on President Reagan. The room fell quiet. The president reached into his pocket and pulled out a card. He began to read from it: "I would rather be ashes than dust; I would rather that my spark should burn out in a brilliant blaze than it should be stifled in dry rot. I would rather be a superb meteor, every atom of me in magnificent glow, than a sleepy and permanent planet."

The president explained that the words came from the American fiction writer Jack London and were once read to Kenny "Snake" Stabler, a notorious but brilliant quarterback for the Oakland Raiders. Stabler was asked what he thought the passage meant. "Throw deep," he replied, referring to the high-risk strategy favored by quarterbacks to drive their teams upfield. With the reading, President Reagan signaled his intention to throw his own support behind the extraordinary machine. On the way back to his office, Trivelpiece was elated. The pitch could not have gone better.

The next day, Trivelpiece took a call in his office. It was John Herrington, the energy secretary, who was preparing for a trip to Switzerland. Reagan had confirmed his support for the supercollider and

Herrington was keen to announce the news, but he wanted to know where the Jack London quote came from before going public.

Within minutes, everyone in Trivelpiece's office was making calls to find out where London had written the passage down. He got people to check at the Princeton Poetry Library and the Huntington Library in San Marino, California, which has a major collection of Jack London manuscripts. He called President Reagan's speech writers, but they didn't know where Reagan had gotten it from.

Trivelpiece knew that Kenny Stabler had written his memoirs in 1986, the year before. He got hold of a copy. It was called *Snake: The On-and-Off-the-Field Exploits of Football's Wildest Renegade*. Trivelpiece pored over every page of the book and couldn't find the quote. "Finally, I put it down and there I saw it, in the preface of the darned thing," Trivelpiece recalls. The passage, it said, had been read to Stabler by *Los Angeles Times* columnist Jack Smith.

Something was wrong though. The Jack London quote in Stabler's book didn't match the president's reading. Stabler had lost the first sentence, but included three more at the end. They read: "The proper function of man is to live, not to exist. I shall not waste my days in trying to prolong them. I shall use my time."

A flurry of calls tracked Kenny Stabler down, but he was unable to resolve the issue. The phrase "throw deep," he added, followed him around, hurled at him from football fans wherever he went. Next, Trivelpiece got Jack Smith on the line. "Those days cost me my wife and my liver," said Smith, suggesting his recollection of events might not be as clear as Trivelpiece hoped. Before ringing off, Smith said there was something Trivelpiece should know. Smith thought the passage was effectively Jack London's communist manifesto. Best not to tell the president, he warned.

In the midst of the excitement, Trivelpiece's assistant, Gretchen Seiler, took a call from her husband. She told him she couldn't talk. They were desperately trying to hunt down a Jack London quote. She read it to him, and her husband recognized it immediately. It was pinned on a bulletin board where he worked.

Soon Trivelpiece had the complete passage: it had been written to an Australian women's rights activist and reported in an article in one of the San Francisco newspapers. It was progress, but still they couldn't give Herrington a reliable reference. Exasperated, Trivelpiece sent a note to Reagan directly, asking him where the passage came from. The reply, when it came, was "a bit harsh," said Trivelpiece. "He thought I was challenging whether or not he got it right, but I just wanted to know where he got it from and why he liked it." Reagan said nothing about the source in his letter. "I never got a clue," Trivelpiece said.

It turned out that Herrington wanted the details of the passage to pass on to George Will, a *Washington Post* columnist, who was writing a piece on the momentous meeting. The next day, the piece appeared under the title "The Gipper Throws Deep."

Back in Edinburgh, Peter Higgs, who was pursuing other work in physics and spending considerable time supervising Ph.D. students, watched events unfold with bemusement. "I think poor old Reagan was thoroughly confused as to whether the machine was going to help him 'zap the commies,' particularly as his science adviser was a former boss at Los Alamos," he said later.

On January 30, 1987, President Reagan's approval of the supercollider was made public. In an accompanying statement, Secretary Herrington described the machine as a "racetrack-shaped particle accelerator" that would be built underground, in a tunnel 52 miles in circumference. That was close in size to the beltway that circles Washington, D.C. When Ernest Lawrence had built his first accelerator more than fifty years earlier, it was no bigger than a shoebox.

Herrington's statement was breathless.[3] The president's decision was of "tremendous significance and historic consequence," it said. "Once again, this nation has said there are no dreams too large, no innovation unimaginable, and no frontiers beyond our reach." He added: "In high-energy physics, the Superconducting Supercollider is the equivalent of putting a man on the Moon."

With one eye on the budget, Herrington made it clear that the government would be looking for cash from other countries to help make the giant collider a reality. In terms of scale and ambition, the project was on a par with President Reagan's other great schemes, including the Strategic Defense Initiative, also known as "Star Wars," with which he hoped to use space-based lasers to cast a protective shield over North America. Like the Star Wars project, the super-collider had more than a scent of exuberance and, to some, no small measure of hubris.

The origins of the supercollider can be traced back to an era long before American physicists were urged to seek "earnest revenge" against Europe in 1983. In the 1950s, Robert Wilson, the former Fermilab director, had dreamed of a particle collider so immense that it would be too costly for any single nation to build. Wilson's "World Laboratory" was visionary, a kind of super-CERN that would be-come the center of human endeavor in the realm of particle physics.[4]

In the aftermath of World War II, the foundations of such a rad-ical global collaboration demanded civility between America and the Soviet Union. In 1959, following talks between Soviet Premier Nikita Khrushchev and President Dwight D. Eisenhower, the heads of the atomic energy organizations of the two nations signed an agreement that paved the way for regular exchanges of scientists and put forward the goal of building a giant particle accelerator.

Six months later, Wilson joined a delegation of physicists to the Soviet Union with the express intention of discussing the construc-tion of a new accelerator. Political events soon derailed any progress the physicists had made. On May 1, 1960, U.S. pilot Francis Gary Powers was shot down over Russia in a high-altitude U-2 spy plane. Eisenhower had been assured by the Central Intelligence Agency that, soaring at 70,000 feet, the U-2 would be safely out of reach of Soviet antiaircraft missiles. The pilots also carried suicide pills and were ordered to set the aircraft to self-destruct if they got into trou-ble, destroying the plane along with its advanced camera and any snapshots the pilot had taken.

The Eisenhower administration gambled on a cover-up story and released news that a weather plane had gone astray and come down in Russia. The truth emerged days later, when Khrushchev showed footage of the aircraft wreckage. The plane was almost entirely intact, and the pilot was alive and well. Tensions between the Cold War adversaries escalated to a new high. The Soviet physicists that Wilson and his delegation were trying to build relationships with could hardly bear to speak to them.

The U-2 incident was followed swiftly by the Cuban missile crisis, which pushed the United States and the Soviet Union to the brink of nuclear war. But in spite of desperate political upheavals throughout the 1960s and 1970s, a small group of scientists from America, the USSR, and CERN continued to sketch out plans for Wilson's World Accelerator. As presidents and premiers fell, Wilson saw the project as a way to heal the wounds of war, to replace suspicion and secrecy with trust and cooperation. "Somehow," he recalled, "in building and operating a World Laboratory, we would not only be exploring nature, but we also might be exploring some of the ingredients of peace."

The machine began to take shape in the minds of the physicists, but at a glacial pace. Leon Lederman formally renamed it the "Very Big Accelerator" (VBA), and the engineers started to look at designs that would drive protons to energies of 10 or possibly even 20 TeV. Toward the end of the 1970s, however, Wilson's dream began to fade. Scientists around the world backed the project, but it was their governments that had to pay for it. By this point, many had already grown impatient and had either embarked on their own accelerators or were heavily invested in CERN.

The election of President Reagan in 1980 set the course for the return of the VBA under a different guise. His science adviser, George Keyworth, had been plucked from the upper ranks of Los Alamos, where he had been a protégé of Edward Teller, the man who had built the hydrogen bomb for President Harry S Truman and who had urged Reagan to embark on the Strategic Defense Initiative. At forty-one years old, Keyworth was young and forceful and saw an urgent need to revitalize American science. He wanted large, exciting projects of the kind Wilson relished, hoping to rid the profession of what he regarded as a growing mediocrity.

Leon Lederman knew an opportunity when he saw one. In 1982, he called on his fellow physicists to back an all-American accelerator called the Desertron, which was based heavily on the VBA designs. At the time, Lederman put a price tag on the machine of $750 million. The Desertron would major in hunting for the Higgs boson, but it would be powerful enough to uncover whatever happens to the laws of nature at energies inconceivable since the Big Bang.

The spur for the Desertron was competition from Europe. For many American physicists, it was bad enough that the Europeans had clinched major prizes with the discovery of the W and Z particles and gluons. What added to American anxieties was a speech made by Herwig Schopper, the director general of CERN, early that year, during his trip to Japan. Schopper had unveiled ambitious plans for the lab that made it clear Europe had no intention of relinquishing its hard-earned place at the top table of physics.

Civil engineers with tunneling equipment were poised to begin work at CERN on a new accelerator, the Large Electron Positron collider, in what was to become the largest construction project in Europe at the time. The LEP collider was a departure from CERN's previous machines in that it crashed together electrons and their antiparticles, positrons. This gave it several advantages over a proton collider, but the most obvious was that the collisions were "cleaner," making it easier to spot signs of new particles created amid the showers of debris. Because protons are built up from other particles—quarks and gluons—the debris kicked out when they collide is far messier.

The LEP accelerator was designed to study the behavior of W and Z particles, but by ramping up to a higher energy, the machine would be a serious contender in the hunt for the Higgs boson. Even before LEP was built, CERN had in mind a formidable upgrade. At the end of its lifetime, which was originally earmarked for the late 1990s, the LEP machine would be stripped from its 17-mile underground tunnel and replaced with an even beefier accelerator. Clad with more than a thousand powerful superconducting magnets, the new machine could contain two beams of particles circulating at 7,000 GeV each, or 99.999999 percent of the speed of light. The upgrade was called the Large Hadron Collider (LHC). A hadron, from the Greek *hadros*, meaning "robust," is any particle made of quarks, such as a proton.

When American particle physicists came together to plan their future at the High-Energy Physics Advisory Panel meeting in the summer of 1983, the concept of the Desertron was ready and waiting. The machine was swiftly repackaged as the more impressive-sounding Superconducting Supercollider, which was designed to accelerate beams of particles to 20,000 GeV apiece, making it more than twice as powerful as the LHC could ever be. In recommending it to government, HEPAP had urged that it be built for less than $2 billion and be ready to run within twelve years. All other proposals for new colliders were dropped. If all went according to plan, the machine would trounce Europe's LHC, rendering it obsolete before it was even built. If all didn't go according to plan, American physicists risked being left behind in the dust.

The Superconducting Supercollider (SSC) became controversial and divisive as soon as it was announced. The plan's unveiling coincided with a call from the Reagan administration to "regain leadership" in high-energy physics, a statement that was seen by many physicists outside America as deeply provocative. The CERN management protested what they interpreted to be an overtly nationalist project that conflicted with their hopes of a coordinated global endeavor. Mathematics suggested that, to be sure of finding the Higgs boson (or whatever else gave particles mass, if it wasn't the Higgs), the next big accelerator must be capable of making particles with energies of around 1,000 GeV, but the SSC risked absorbing vast amounts of public money that could be spent more shrewdly. CERN wanted to see the LHC built on its own site and a large and complementary linear electron-positron collider built in the United States. The LHC could be built at CERN for a fifth of the cost of the SSC, they argued, not least because the tunnel for it already existed.

International protests came from further afield than CERN. Nations involved with Wilson and Lederman's plans for the VBA took offense that decades of work on the machine had fed into the blueprints of the SSC and were effectively being subverted to reestablish America as the world leader in physics. Speaking to the House of Representatives Science and Technology Committee in February 1985, President Reagan's science adviser, George Keyworth, said: "The nation, or group of nations, that builds the SSC will become the new world center in high-energy physics. I won't conceal my opinion that it would be a serious blow to US scientific leadership if that facility were built in another country."

At CERN, Carlo Rubbia was given the job of making the LHC a reality. Rubbia, who was awarded the Nobel Prize in Physics in 1984 with Simon van der Meer, the accelerator engineer, for their work leading to the discovery of the W and Z particles, was appointed head of a new committee to plan the future of the lab. Only weeks later, as Keyworth was pressing home the importance of making the SSC an all-American machine, Rubbia flew to Japan, where he told physicists not only that CERN could build the LHC by 1992, but that it could be completed on a budget of only $250 million.

The move, some said, was classic Rubbia. It was bullish, optimistic, and ambitious. It enticed Japanese physicists—and others looking on—with the tantalizing prospect of playing a serious role in a machine that could bag the best discoveries before the SSC was even up and running. Lederman could only raise an eyebrow at Rubbia's numbers. He found it bizarre that CERN was claiming it could build the LHC so quickly and for such a small amount of money.[5] He wondered if Rubbia was talking in terms of 1940 dollars. Writing in *Discover* magazine in July 1985, the U.S. journalist Gary Taubes claimed that "some scientists suspect that [Rubbia's] dubious numbers are bluffs to force the US into collaboration [with CERN]."

Soon, American particle physicists were facing criticisms closer to home from researchers working in other areas of physics. The Department of Energy proposed the SSC on the promise that it would be built with "new money" and support from outside nations. But soon after the project was announced, $18 million was diverted from the scrapped Isabelle collider project at Brookhaven National Laboratory to get work started on the SSC. Physicists working in other areas feared that funding for their own research would also be raided to pay for the flagship collider.

The SSC was the largest pure science project ever, and as such it was inevitable that it would face major hurdles along the way. The management of a project of such scope was only one difficulty. There were also disagreements over the machine's capabilities and over designs for the powerful superconducting magnets that were critical to its operation. While none of these proved insurmountable, periodic reviews of the budget saw the projected costs rise steadily. At a time when the U.S. budget deficit weighed heavily on Congress's shoulders, the soaring costs were an ominous sign.

In the months after President Reagan formally endorsed the new accelerator, leading advocates and proponents of the project were invited to give evidence at a series of hearings held by subcommittees of the House of Representatives. One of the leading critics was the Nobel Prize–winning physicist Philip Anderson from Princeton University, who had suggested the idea of a Higgs-like mechanism a year or so before Peter Higgs and others wrote the theory out in full. Anderson, who worked on superconductivity and magnetic effects in materials, had a history of opposing large amounts of public expenditure on high-energy physics. When particle physicists argued that their work focused on more fundamental questions of science than his, Anderson argued that it was no more fundamental than the work Alan Turing had done for computer science or that James Watson and Francis Crick had achieved with unraveling the structure of DNA.

Another leading figure who testified against the supercollider was James Krumhansl, a distinguished materials scientist at Cornell University. Krumhansl's testimony carried particular weight at the time because he was lined up to take over the presidency of the American Physical Society, the nationwide organization that represented the field of physics. Krumhansl argued that the supercollider was likely to harm the wider pursuit of physics by drawing money away from promising research. A colleague of Krumhansl's, James P. Sethna, later explained that Krumhansl "valued the science of the supercollider highly, but he did not value it a thousand times more than the other fields of science."[6]

Krumhansl felt the machine was needlessly large and expensive because it relied on old technology. A year earlier, physicists had discovered new kinds of superconducting materials that didn't need to be kept as cold as the conventional sort, which had to be bathed constantly in liquid helium. With the new materials, engineers could build much stronger and cheaper magnets, or so the argument went. The fact that it had taken twenty years to turn the first superconductors into useful accelerator magnets was pushed to one side. "They unquestionably have the potential to save billions of dollars in construction and operation of particle accelerators like the SSC," Krumhansl wrote.[7] "I have little hesitation in predicting that they will be brought to technological usability in three to five years." Some congressmen claimed that the collider would need only 1 percent of the land area if it was built with the new magnets. The supporters of the supercollider saw this as a delaying tactic, a reason to put the project on ice until the new materials became available.

Steven Weinberg, whose 1967 theory had unified electromagnetism with the weak force using the Higgs mechanism, was invited to argue the case for building the extraordinary collider. He didn't relish the thought of testifying before Congress and knew he couldn't disagree with some of the arguments Anderson and Krumhansl would make. Writing in his 1993 book *Dreams of a Final Theory*, he recalled: "All this time it had been a nightmare of mine that I would be called up before some tribunal and asked in a stern voice why it is worth $4.4 billion to find the Higgs particle." If the money was spent on the SSC, perhaps it wouldn't feed into new gadgets and technologies as quickly as if it were spent on more applied physics. Nor was he sure that particle physics was intellectually more profound than some other fields of science. But he felt that, in some way, the search to understand the basic building blocks of nature and the forces that acted upon them was more fundamental than other areas of the science.

Weinberg told the committee that particle physics was unveiling something about the structure of the universe at the deepest level. The laws of nature, he said, seemed to become more coherent and universal the better they were understood: "We are beginning to suspect that this isn't an accident, that it isn't just an accident of the particular problems that we have chosen to study at this moment in the history of physics, but that there is simplicity, a beauty, that we are finding in the rules that govern matter that mirrors something that is built into the logical structure of the universe at a very deep level."

When Weinberg finished, a Republican congressman, Harris Fawell of Illinois, thanked him and the other scientists for their comments, adding that he wished there was one word to sum up the point of building the supercollider. He turned to Weinberg and said, "You said you suspect that it isn't an accident that there are rules which govern matter and I jotted down, Will this make us find God? I'm sure you didn't make that claim, but it certainly will enable us to understand so much more about the universe?" At that point, Don Ritter, a Republican from Pennsylvania, who had looked unfavorably on the project, interjected, "If this machine does that, I am going to come round and support it."

Weinberg kept quiet. It was best the congressmen didn't hear his views on the prospects of the supercollider enabling physicists to find God. Had he ventured his opinion, it wouldn't have helped the project.

Another distinguished physicist to give evidence in support of the supercollider was Burton Richter, director of the Stanford Linear Accelerator Center. After his testimony, he was asked a question that mirrored Harris Fawell's thoughts. Tim Valentine, a Democrat from North Carolina, quizzed Richter on the value to humanity of understanding the constituents of matter and the forces that govern them.

"You ask tough questions," Richter replied. He then set out his thoughts:

> You'd know how the universe was born, you'd know how it evolved, you'd know where it's going to be now and in the future. You'd know everything there is to know about our physical world, and you would know much better what man's place is in our physical world.
>
> On a more practical level, if you knew all about it, maybe you could control it better. Always in the past when we have learned more about the physical world, we have been able to control it to do good things or bad. We learned 150 years ago . . . about electricity and magnetism. . . . From that knowledge about the physical world has come lights and television and what-have-you. I don't think that anybody would promise that in the future we can control it better, but always in the past, if we learned that, well, I would say we have found out it is true that knowledge is power.

Richter's comment was a reminder of how understanding matter at the deepest level had given society the technology to transform and flourish, on the one hand, or bomb itself back to the Stone Age, on the other. But it also cast forward to the prospect of future technologies built on whatever physicists found next. Few scientists today openly speculate on the possibility of controlling the Higgs field—because of the vast amount of energy that would be needed to manipulate it—but those who do warn that tinkering could have extraordinary implications. Altering the Higgs field—a surely impossible job, given that it would probably involve heating space to one million billion degrees Celsius—would be so disastrous that we would not be around to witness it. Lumps of matter, from people to planets, could become unstable as their constituent particles lost mass and vanished in a puff.

In Washington, D.C., Alvin Trivelpiece was getting it in the neck. Buoyed by President Reagan's support for the SSC, scientists were dreaming up tweaks to improve the machine's design, but the requests for this whistle and that bell did not go down well.

"The ink was hardly dry on the deal when I started getting calls from physicists saying now it was agreed, we really should have a bigger aperture, or why don't we have a bigger injector," Trivelpiece recalls. "I'd told congressmen and senators this was an experienced group, that they can deliver on time and on budget, but then they started saying the cost will be $5.5 billion or $6 billion. I got calls from Congress saying I'd basically lied to them; they said the physicists had lied when they talked me into it. It all got a bit ugly."

Then there was the question of where to build the machine. While arguments raged over the wisdom of even funding the supercollider, Trivelpiece arranged for Robert M. White, president of the National Academy of Engineering, to begin looking for a suitable home for the accelerator. By the time the deadline arrived, the Department of Energy had forty-three proposals from twenty-five states, weighing a combined three tons. Weinberg was on the site selection committee. More than twenty years later, boxes of proposals from hopeful states are still piled up in his office in Austin.

States rightly saw the supercollider as a ticket to economic regeneration. In some cases, they were clearly desperate. The successful site needed solid political support, good roads and other infrastructure, and crucially, a nearby airport to ferry visiting scientists to and fro. A history of flooding or earthquakes would immediately rule a site out. One proposal, from Nevada, was pitched as being only half an hour from Reno airport. Half an hour by air, that is. "You'd have to fly to Reno, then take a private plane to get to the site," Weinberg said. "A non-starter. Ridiculous."

The sites were gradually whittled down to a short list of seven, leaving only Arizona, Colorado, Lederman's Fermilab site in Illinois,

Michigan, North Carolina, Tennessee, and Texas still in the race. As other states heard they were out of the running, their support for the project also fell by the wayside.

Two proposals stood out. The Illinois site was home to Fermilab, which was accessible, offered all the amenities that visitors and resident scientists might need, and already bustling with highly experienced scientists and engineers. The other state was Texas, which won out on local support and geology—it lay on a bed of easily drilled chalk and had already raised $1 billion to help fund construction.

On November 10, 1988, the winner was announced. The accelerator would be built in Ellis County, Texas. At Fermilab, Lederman called an all-staff meeting to break the news, though many had already heard it over the radio while driving to work. The mood was gloomy. The rise of the supercollider would ultimately mean the death of Fermilab. Two days earlier, George H. W. Bush of Texas had been elected president of the United States. Inevitably, there were suspicions of political interference, but Weinberg, for one, insists politics did not figure into the decision.

President Bush supported the supercollider and encouraged other nations to take on a share of the costs. By this time, Carlo Rubbia had risen to director general at CERN, and he was looking for a commitment from America to help build the LHC in return for help from Europe in building the SSC. The proposal came to nothing more than talk. Keyworth said that directing funds to the LHC as well as the SSC would break the bank.

In 1991, civil engineers finally began construction on the supercollider. Magnet development and testing labs went up. So did housing for the enormous refrigeration units needed to circulate liquid helium at −196 degrees Celsius around the machine's thousands of superconducting magnets. Underground, mile upon mile of chalk was carved away to make room for the accelerator. The cost of the project had risen to $8 billion.

Work went ahead on the basis of it costing the government no more than $5 billion, with the rest coming from foreign nations.

Securing a hefty contribution from Japan was a top priority when President Bush visited that country at the beginning of January 1992. The trip was not overwhelmingly successful, and President Bush left the country empty-handed. Japanese government officials said the SSC was not an international project and wouldn't get their backing until it was.[8]

That summer, the U.S. House of Representatives voted to cut the nation's losses on the SSC, but the decision was reversed by the Senate. The next June, when Bill Clinton was six months into his presidency, the same thing happened again. By now, the General Accounting Office was estimating that the final cost of the SSC would reach $11 billion.

In September 1993, a distinguished delegation of physicists, including Steven Weinberg, Burton Richter, and Leon Lederman, left their labs and offices for George Washington University to promote the supercollider. Prominent British physicist Stephen Hawking sent his own message of support via videotape. They hoped for widespread media coverage, but the story sank beneath the main news of the day: Clinton had brought the Israeli prime minister, Yitzhak Rabin, and PLO chairman Yasser Arafat together to sign the Oslo peace accords.

A month later, Congress was scheduled to vote again on Big Science projects. The SSC was up for approval again; so, too, was the International Space Station, a multinational venture with an initial price tag of $25 billion. The space station survived, its funding secured by just one vote. Voting on the supercollider took place the next day, and to the dismay of its supporters, the vote went 2–1 against the machine. The supercollider was scrapped just as engineers had completed more than a quarter of the tunnel beneath the Texas countryside. Some 2,000 people had taken jobs on the collider, and $2 billion had already been spent.

The death of the supercollider, the first dedicated Higgs-hunting machine, was the end of an American dream for many particle physicists. The thousands who had given up hard-earned faculty positions to relocate to Texas saw the project vanish like a mirage before their eyes. Their careers had reached a cliff edge. The supercollider was conceived to make American particle physics world class, but, by rejecting the project, Congress risked condemning the field to a future of mediocrity.

The demise of the supercollider felt like a fatal blow to many people following the hunt for the Higgs particle, but at least part of that impression came from it being sold so strongly as a single-issue Higgs-hunting machine. Notes on marketing the machine drawn up by Lederman at the time state: "After much trial and much error, the thing that works best (lay audiences, other sciences) is to pick on the symmetry breaking: i.e. the Higgs." Although the supercollider project failed, American physicists still had the Tevatron, which had switched on two years before Alvin Trivelpiece had pitched the SSC to President Reagan. The Higgs particle might be at the very extent of the Tevatron's reach, but with a substantial detector upgrade and other improvements, the Illinois machine became America's best hope of discovering the particle.

The supercollider was a costly debacle and a low point in the story of the hunt for the Higgs boson. As a case study of failure, it raises serious questions about government priorities, the political savvy of scientists, and the value to society of high-energy physics.

Weinberg admits he is still bitter over the supercollider affair. He thinks Krumhansl's opposition was "disgraceful" because, as president-in-waiting of the American Physical Society, his voice carried disproportionate weight. More importantly, he regrets that Texas won the site selection. The decision, in the end, was between Texas and Lederman's Fermilab site in Illinois. "The people in Chicago were all 'not in my back yard,' and that was a big factor. It may have survived if it had gone to Illinois," Weinberg said. "I wish we'd chosen Illinois."

When it was announced that the machine would be built in the Texan town of Waxahachie, the supercollider immediately, though unexpectedly, found itself in competition with the International

Space Station. The scientists backing the machine had not foreseen this. The supercollider was designed to claim scientific superiority by probing nature in a tunnel deep underground. The space station was a show of technological prowess, flying 200 miles high over the Earth's surface. Though wildly different in so many ways, both were earmarked as Texas projects, the space station being run from the National Aeronautics and Space Administration's Johnson Space Center in Houston. Many members of Congress balked at the thought of voting for two massive projects in Texas. Of the two, there were simply more votes in the space station, which had defense-industry backing.

Bill Clinton's election as president didn't help the project's chances of surviving. Though not against the supercollider, Clinton was less invested in it than either Reagan, who advocated the machine from the start, or President George H. W. Bush, who had close connections with Texas. Clinton brought in a utilities company manager, Hazel O'Leary, as energy secretary, and O'Leary did little to fight for the machine. Some scientists claim that Clinton told the then Texas governor, Ann Richards, that his administration would not support both the SSC and the space station, and let her choose which one to save.[9]

Alvin Trivelpiece still works as a consultant at government labs, though he is officially retired these days. He says there is "no end to the sadness" he feels for those whose dream was shattered by the termination of the SSC. People who helped bring down the supercollider moan that America has too few scientists and engineers. "They don't get it," he says. "There's a connection between having inspiring goals and great career paths. It's an American tragedy."

Trivelpiece can't help wondering whether the supercollider might be running today had scientists agreed to name it after President Reagan, a suggestion doing the rounds in the early days of the machine's story. The supercollider would have transformed into the Reagan Accelerator Center. There was even talk of a dedication ceremony with five living presidents. "Can you imagine having the funding for the accelerator turned off if five living presidents had come down and put the show on the road and agreed to go forward with it? There would have been nothing political that could ever have turned it off!" he says.

The supercollider was not felled by one massive blow but by a barrage of events that gradually undermined it. "In Washington, it's not necessary to kill something off. All that's necessary is to not fight for it like a tiger, and it will die on its own. With the Superconducting Supercollider, a thousand cuts caused it to bleed to death," says Trivelpiece.

The year the supercollider died was the year the "God particle" was born.[10] Leon Lederman and Dick Teresi, an American science writer, copublished a history of particle physics that set the stage for the supercollider's hunt for the Higgs boson. According to Lederman, the book's editor rejected any titles that mentioned Higgs and his mysterious boson. They had to be more inventive. Lederman claims they wanted to call the book *The Goddamned Particle* because it was so hard to find, but instead they settled for *The God Particle*. The nickname was well deserved, Lederman wrote, because the particle is critical to our understanding of matter, yet deeply elusive.

The nickname ranks as one of the most, if not the most, derided in the history of physics. Working scientists rant aloud about how profoundly dreadful a name it is. Some think it is plain lame. Some hate the fact that the media embraced the name for no better reason than that it sounds more intriguing than the "Higgs boson." Peter Higgs winces at it. He worries it is aggrandizing, even offensive to those with religious beliefs.

Lederman is approaching his nineties now and lives in a house on the grounds of Fermilab. He admits to receiving a barrage of complaints for coming up with the nickname, and he once quipped that the name offended two groups of people: those who believe in God and those who don't. The backlash is easy to understand: the nickname is meaningless and suggests that physics can lift a veil on spiritual matters.

Lederman is adamant that if the supercollider had been built, physicists would have discovered the Higgs boson long ago. "We would have found the Higgs boson by 1998 or 1999," he says. "We would have either found it or said there's something else going on with the whole Standard Model."

As the fortunes of the supercollider rose and fell, engineers in Europe completed construction of their own enormous particle accelerator. Though smaller than the visionary supercollider, the Large Electron Positron collider occupied a tunnel 17 miles in circumference. Buried 100 meters beneath the ground at CERN's lab near Geneva, it was designed to create and study Z particles before pushing on to higher-energy territory where the Higgs boson was thought to be hiding.

The story of the LEP collider begins long before the machine started crashing particles together in 1989. The construction of the machine, the largest particle accelerator in the world, was an adventure in its own right. For scientists hunting for the Higgs particle, it seemed nothing was ever simple.

7

Massive Maggie

From Geneva International Airport it is a short drive west along Route de Meyrin to the main CERN complex. The campus covers 30 hectares all told and sits directly over the French-Swiss border. At the entrance, security guards ask for identification, disappear for a moment, and return to wave people in. Beyond the barrier is a sprawling set of office blocks and car parks linked by looping roads and walkways.

CERN is a world-class laboratory that is almost duty-bound to look plain to the casual observer. The organization receives subscriptions from twenty member states worth a billion dollars a year, but none of it is earmarked for imposing buildings and manicured gardens. At CERN, the real money is spent underground.

Few people know more about what CERN does with its budget than Lyn Evans. In forty years of service at CERN, he has worked on practically every particle collider the laboratory has built. You could say working underground is in Evans's blood. He was born in the Welsh coal-mining village of Aberdare to a father who spent most of his working life down the pit.

Evans was in charge of the Large Electron Positron collider, which was constructed at the European lab between 1983 and 1988. With the demise of the Superconducting Supercollider in the United States, LEP physicists became the first particle-accelerator scientists to begin a serious hunt for the Higgs boson.[1] Although the Tevatron was up and running at Fermilab in 1985—four years before LEP saw its first collisions—it had not yet smashed together enough particles at the energies necessary to reveal the Higgs. It would take a major refurbishment of the machine and its detectors in the late 1990s for the Tevatron to get back in the running to discover the particle.

When LEP was built, it was the largest scientific instrument in the world and the most sophisticated CERN had ever attempted. Building and operating it brought up enough problems to keep engineering students in case studies for a decade.

CERN had built its reputation by colliding protons together, but LEP was to depart from that tradition. The machine was designed to crash electrons into positrons, their antimatter equivalents. Because these are truly fundamental particles—in that they aren't made up of smaller parts—slamming them headlong into one another produces a pure burst of energy. In a flash, that energy becomes freshly minted matter.

The LEP collider was designed to run in two stages. In the first, beams of particles were accelerated to around 50 GeV apiece before colliding head-on. The objective was to make vast numbers of Z particles so that they could be studied down to the minutest detail. In the second stage, the beams would be ramped up to 80 GeV each, enough to produce pairs of W particles, so they, too, could be studied intimately.

If you want to build a particle accelerator like LEP, the factors you need to consider are endless. Before anything, you face a trade-off. Small ring-like colliders are inefficient because when you accelerate electrons in a tight circle they shed a lot of energy in the form of radiation.[2] Larger accelerators are more efficient because the electrons corner more gently, but they cost far more to build.

CERN's early plans for LEP involved constructing a 50-kilometer ring, but the estimated cost was sky high. The organization settled on a 27-kilometer ring, 4 meters in diameter, that would be fitted with four huge detectors to look for new matter that might flicker into being with every collision.

Once you've decided on the size of your machine, you have to find somewhere to put it. The decision isn't as easy as you might think. Buying up enough land to build it next door to CERN's main site would have broken the bank. It was cheaper to bring in heavy-duty boring machines and make room for the thing underground.

Going underground has a lot going for it. The Earth's crust is an effective shield against the radiation given off by the machine. It makes it easier to protect from anyone who might want to sabotage it. And it saves the organization from having to construct a metal monstrosity enclosed by a barbed wire fence across what is otherwise beautiful rural farmland.

There are downsides, too. Constructing a machine of LEP's size underground takes meticulous planning and ingenious engineering. And when things go wrong, making repairs is exacting work.

CERN opted to go subterranean, but the decision created a whole new list of problems. Above ground, the LEP ring was too large to squeeze between the Geneva airport on one side and the Jura Mountains on the other. Underground, things weren't much better. The airport was built on a loose mix of rock and soil that was nightmarish to tunnel through. Deep beneath the Jura range, the limestone was riddled with faults and cracks that were filled with water. A geological report on the site warned that trying to cross just one of these faults could add $16 million to the bill and delay the project by more than a year.

Before pushing ahead, CERN engineers drilled enough exploratory boreholes for scientists to put together the most comprehensive geological survey of the region ever conducted. With it, they could finally work out what to do. After shifting the position of the tunnel slightly, they tilted it so that it ran gently downhill, from around 50 meters deep at the Jura foothills to more than 100 meters deep at the airport. That way, almost all of the concrete-lined tunnel would pass through solid rock.

Under the local laws around CERN, you can't just go and build a whopping great particle accelerator—or anything else for that matter—under someone's house without asking them. The LEP collider crossed the jagged French-Swiss border four times and went directly under homes on both sides. If you lived in Switzerland, the law said you owned 30 to 50 meters of rock beneath your house. The authorities wanted to prevent utility companies from ruining residents' plans for underground garages or even for sinking artesian wells. In France, the law was different and in a fabulous way. If you lived in France, you owned the chunk of ground your house was on all the way down to the center of the Earth.³ The collider tunnel was too deep to worry the Swiss, but, in France, the CERN management needed to get written permission from around 2,000 homeowners before they could start digging. Several of them refused at first, and it was two years before a settlement was reached between these holdouts and the French government.

CERN's relations with the nearby rural farming community were not improved when a physicist who had worked at the lab warned that LEP would produce as much pollution as a million cars and destroy their harvests in the process. Unsurprisingly, the farmers and local authorities were in an uproar. The physicist had raised a serious point but mangled his equations. The particles whizzing around inside the LEP accelerator would release intense X-rays, and if these got into the air, they could produce toxic gases such as ozone and nitric oxides. Above ground, these would have been problematic, but LEP was underground and shielded with lead, aluminum, and the magnets that steered and focused its beams. When CERN took this into account, they worked out that the pollution rising from LEP's ventilation shafts was the equivalent of just a few extra cars on the local roads.

When construction finally began in 1983, LEP became the largest civil engineering project in Europe. The 27-kilometer tunnel was just the start. Vertical access shafts were excavated at eighteen points around the tunnel so that people could get in and out. Four enormous caverns were cut from the rock around the tunnel to make room for the collider's detectors. A further sixty alcoves and chambers were added. In the end, less than half of the 1.4 million cubic meters of rock pulled out of the ground came from the main tunnel.

CERN planned every step of the LEP project thoroughly but despite all their precautions, accidents happened. In September 1986, the boring machines hit a geological fault and flooded the tunnel with water. It took eight months to stem the flow, drain the tunnel, and get back to excavating. Near the airport, the crew hit another snag while trying to bore three vertical shafts. The soil and rock beneath the surface was saturated with water down to at least 100 meters, forming a natural reservoir for the local population. To get around the problem, engineers forced tubes into the ground where the shaft was supposed to go and pumped them full of coolant. This froze the ground into a solid block that could be excavated as normal.

The LEP tunnel was completed a week before Valentine's Day in 1988. Thanks to the satellite- and laser-guided boring machines, when the two ends of the tunnel met, they were off by just 1 centimeter. The construction work alone had taken six years.

Later that year, before LEP switched on, the British prime minister, Margaret Thatcher, gave a speech at the Royal Society in London, the most prestigious scientific organization in the country. She spoke wistfully of Arthur Eddington, whose stories Peter Higgs had read as a schoolboy. Eddington had chosen the Royal Society as the venue in which to describe his 1919 expedition to the west coast of Africa to prove Einstein right by watching starlight bend around the sun. "When Arthur Eddington presented his results to this society in 1919 . . . it made headlines," Thatcher said. "Many people could not get into the meeting, so anxious were the crowds to find out whether the intellectual paradox of curved space had really been demonstrated. Should we be doing more to explain why we are looking for the Higgs boson at CERN?"[4]

The story Thatcher was referring to was so big that the *New York Times* had given it not one headline but six. The article appeared on November 10, 1919, by special cable from London under the headline

"Lights All Askew in the Heavens." Beneath that, the newspaper declared: "Men of Science More or Less Agog over Results of Eclipse Observations." In case anyone might fear the sky was falling, it clarified further down: "Stars Not Where They Seemed or Were Calculated to Be but Nobody Need Worry."

The LEP collider lurched into action in August 1989. The first job CERN physicists set out to do with their new particle collider was to measure the mass of the Z particle more precisely than ever. To do that, they needed to know precisely how much energy was in the particles flying around inside the machine. Nail that down, and they would know exactly how much energy had gone into each collision that created a Z particle. From that, they could deduce its mass.

After two years of taking data, in 1991, something odd happened at LEP. Strange patterns were spotted in CERN's measurements that showed its beams growing stronger and then weaker with baffling regularity. Evans, who was in charge of the machine, suspected dodgy equipment was to blame—perhaps a misbehaving power supply.

Scientists at CERN scratched their heads over the strange signals for months and, in time, word got around to researchers in other laboratories. One day, Gerhard Fischer, a scientist at the Stanford Linear Accelerator Center, called CERN with a hunch. He suggested the European lab try something.

To follow up the tip, Albert Hofmann, an old hand at CERN, conducted a long and tiring experiment that ran from midnight until 4 A.M. the next morning. The experiment—and what it found—has become legendary at CERN. There among the data Hofmann recorded was the culprit Fischer had suspected: our nearest celestial neighbor, the moon.

At school, everyone learns how the gravitational tug of the sun and the moon give rise to tides on Earth. Although the Moon is only a fraction of the mass of the sun, it has a greater effect on tides because it is so much closer to us.

What is less well known is that the moon and the sun also create tides in the Earth's crust. Hofmann's experiments showed that when the sun and moon aligned, they produced bulges in the ground that made the surface at CERN go up and down by around 25 centimeters.

The ground was shifting—literally—beneath LEP's feet.[5] As the ground swelled with Earth tides, the LEP ring stretched by roughly 1 millimeter. The change, imperceptible to anyone working on the machine, altered how far particles inside the collider had to travel in each lap. The effect was minuscule, but enough to change how much energy the particles carried. "The machine was exquisite," says Evans. "To think we could get that kind of precision with such a huge machine is just amazing." The inside joke at CERN—and it was only half joking—was that scientists better buy themselves an almanac, so they could check where the sun and moon were while their experiments were running.

As work on LEP carried on, it became clear that the sun and moon weren't the only things interfering with the machine's operation. Evans's team corrected for the Earth tides, but then spotted even more subtle fluctuations in the energies of the beams. The new signals were even more puzzling. From Monday to Friday, the energy of LEP's beams wobbled at exactly the same time every day. They wobbled on the weekend as well, but at a different time. The bizarre effect was so predictable you could set your watch by it.

Scientific collaboration doesn't get more international than at CERN. Twenty countries have a direct hand in running the lab, but many other nations are involved. The scientists working on experiments there come from nearly six hundred institutions around the world. Many come and go through Gare de Cornavin, the main train station in Geneva, and it was here that CERN scientists found the source of the latest peculiar effects to plague their machine.

If you were a regular at the Geneva railway station, the timing of the wobbles in LEP's beams might have seemed vaguely familiar. The high-speed Train à Grande Vitesse, or TGV, that links Geneva with Paris is a punctual beast, and the fluctuations at CERN coincided to the minute with the train's departure from the station.

CERN gathered together a small team to investigate how the TGV was interfering with the collider. They put sensors around the LEP machine and took readings of electrical currents at Cornavin and other railway stations. Soon, they pieced together what was happening. When the TGV was ready to leave Cornavin station, it needed a surge of electrical current to power itself up. Some of that current passed into the railway tracks and leaked into the ground. The CERN team worked out that the stray currents were finding their way into the LEP ring, boosting the machine's magnetic fields and raising the beam energies by up to 12 MeV.

The strange case of the TGV and the phases of the moon put CERN scientists on their guard for other curious signals that might wander uninvited into their experiments. Before the machine was retired for good, LEP had recorded changes in the water level of Lake Geneva and picked up an earthquake in Turkey more than 2,000 kilometers away.

As soon as LEP began colliding particles, scientists were on the lookout for signs of the Higgs boson. The Higgs could crop up in a variety of ways, but the easiest time to spot a possible Higgs—early on at least—was when freshly made Z particles collapsed into a clutch of particles, including what might be Higgs bosons. The teams on LEP's four detectors knew what these particle tracks should look like, but equally, they knew that unless the Higgs particle was incredibly light, it could be years before a telltale Higgs signal revealed itself for certain.

A few months after the machine was switched on, a scientist involved in the Higgs search, Jean-François Grivaz, gave a half-hour talk in the CERN auditorium. He walked people through the different ways the Higgs might appear, but not much else: after all, the machine had only produced a few weeks' worth of data, and it was far too early to have seen anything. When the half hour was up, Grivaz thanked his audience and invited questions. One hand slowly reached into the air. It was Jack Steinberger, a CERN physicist who had shared the Nobel Prize with Leon Lederman the year before. "I was asleep for most of this," Steinberger said before getting to his question. "Did you find the Higgs?" he asked.[6] The room erupted with laughter.

CERN spent billions of Swiss francs building LEP in the hope of learning more about W and Z particles and hunting for the Higgs particle. The machine had some teething troubles, but so does any ambitious project. Once completed, the machine was so sensitive it told scientists more about the world around them than they'd ever expected to learn. At the very least, it detected how much rain had fallen over Lake Geneva and how much the water levels would rise, got them to pay attention to when the next full moon would be in the sky, and taught them when the last train left for Paris.

D avid Miller likes to sing baritone and popularize science, and on occasion he does both at the same time. He also lays claim to being in one of the most unruly mathematics classes in 1960s London, one that happened to be taught by Peter Higgs at University College.

After his stint at the university, Miller spent time at Brookhaven National Laboratory in New York and at Fermilab near Chicago, but later on he returned to Europe and took a job at CERN. When Miller arrived at the laboratory, the LEP collider had been up and running for a while and had produced some beautiful results. What it had not produced—as far as anyone could tell—was any real evidence of a Higgs boson.

One afternoon, Miller left his office at the lab and climbed aboard a shuttle bus that had pulled up to take people to Geneva airport. As he sat waiting, the bus filled up with science journalists who were on a visit to find out what physicists at the laboratory were doing. Before long, he fell into conversation with them. "They said they had been to see this fellow, John Ellis, who had been trying to explain to them what the Higgs boson was," Miller recalls. "They hadn't understood a word of it."

Exasperated, the journalists begged Miller to explain the Higgs boson in the simplest language he could find. Miller sat and thought for a moment. He wasn't sure he could make them grasp it. Then he had an idea. "Imagine all you men are in a room and you're arguing about some story or other, when all of a sudden a very beautiful woman walks in." He paused to let them imagine. "As she walks through the room, those of you nearby forget what you were arguing about and form a cluster around her. By gathering around her, you impede her progress, you slow her down. It's as if she's become heavier." The analogy had flaws, but the journalists got it. The room of reporters represented the Higgs field. The woman was a particle that gained mass by interacting with the field. And the cluster of fawning men was the Higgs boson.

While LEP was running, CERN had one eye on the machine's successor, the Large Hadron Collider, which they hoped to build in the same underground tunnel. Before the project could be approved, CERN needed to be sure it had the backing of the countries that funded the lab. But the request for support came at a bad time for the British government. Margaret Thatcher had just left office and had been replaced by John Major, who was looking to make deep cuts in the physics budget. Britain's continued involvement in CERN was in the balance.

In April 1993, Britain's Institute of Physics held its annual conference in Brighton on the south coast. The talk everyone wanted to hear came from William Waldegrave, the science minister. Echoing the words of his former boss at the Royal Society in 1989, Margaret Thatcher, Waldegrave told scientists they had to get better at explaining their work to the public if they wanted to keep the government's support.

Waldegrave set the physicists a challenge. CERN was costing Britain a huge amount of money, but he had no idea what its main quarry, the Higgs boson, was or why it mattered. To be fair, hardly anyone else who wasn't a physicist did either. The challenge was to explain in plain English what the Higgs was and why it was so important. "If you can make me understand that, I stand a better chance of helping you get the money to find it," Waldegrave told the crowd. The winner, he promised, would be rewarded with a bottle of vintage champagne—paid for out of his own pocket.

That week, the editor of the prestigious British science journal *Nature* devoted a whole page to the competition. John Maddox was a legendary and much-loved figure in British science. Among other things, he was known for settling down with a bottle of wine and a packet of cigarettes before bashing out his editorials, often the night before the journal went to press. His piece on the Waldegrave challenge attempted to give readers some tips on how to win the competition, but more intriguing were the final two sentences of the article. They read: "The real interest of the hunt for the Higgs particle is that it may not be quite what is expected. But best not to say that too loudly."[7]

At his home in London, David Miller had heard all about the Waldegrave challenge. It reminded him of the time he was leaving CERN and was asked to explain the Higgs particle to a group of journalists. He thought the explanation he had come up with was good, maybe even good enough to win himself a bottle of vintage champagne. For a second opinion, he ran the analogy of the beautiful woman and the press pack past his wife. She hit the roof. It was sexist, she said. Appalling! He mustn't send it off to the science minister. Miller had other ideas though.

Five months later, Waldegrave finally announced the results of the competition. At the time, the British government knew it could hardly afford to pay its annual subscription to CERN—£55 million a year—and cover the extra costs of taking part in experiments planned for LEP's successor, the Large Hadron Collider. Waldegrave said the entries—there were 117 of them in all—made him appreciate the importance of finding the Higgs particle. He told physicists: "If we cannot find the money—and that is going to be hard-pounding—I will recognise it as a loss."

There were five winners in all, but no one doubted which was the best. It came from a Dr. David Miller at University College London.

It went something like this: "Imagine a cocktail party full of MPs [members of Parliament] who are evenly distributed around the room, all talking to their nearest neighbors. All of a sudden, a woman appears at the door. When the politicians look over, they realise it is none other than the former prime minister herself, Margaret Thatcher." Miller didn't change the rest of the analogy, and Waldegrave loved it. Miller's winning entry was quickly picked up around the world and it has become one of the most popular ways of describing the Higgs field and its particle. If it wasn't for the wise judgment of his wife, the analogy might never have seen the light of day.

The winning explanations made for a perfect introduction to the particle that physicists were so desperately hunting for. But what Waldegrave didn't learn from the winners' entries was how, over the years, scientists had refined the role of the Higgs field. Their work clarified how much mass the Higgs field was responsible for, and also how another field, which behaves just like the Higgs field, played a crucial role in "inflation," a critical period in the earliest moments of the universe when its size expanded at a fantastic rate.

Frank Wilczek probably knows more about the origin of mass than anyone else. Of Polish Italian descent, he was born and raised in Queens, New York, and joined Princeton University as a graduate student in 1971. The work he began there—at the tender age of twenty-one—earned him a Nobel Prize more than thirty years later.

Wilczek, who is now at the Massachusetts Institute of Technology in Cambridge, Massachusetts, shared the award for discovering something curious about the quarks that make up protons and neutrons inside atomic nuclei. He showed that the further the quarks were pulled apart, the more strongly they were drawn to one another; and when they were very close together, they behaved as if they were essentially free. It was as if they were connected by elastic bands. Normal forces don't behave that way: they usually get weaker with distance. Wilczek's work has since become a cornerstone of the Standard Model, the rules that govern how particles in nature behave.

Wilczek's work has far-reaching implications, especially when it comes to understanding where mass comes from. While the Higgs field is thought to be responsible for giving mass to quarks and electrons, the masses of these individual subatomic building blocks contribute almost nothing to the mass of a single atom. Nearly all of an atom's mass comes, not from the cumulative weights of quarks and electrons inside it, but from the energy stored in the quarks as they move around and in the field that binds them together.

At first glance, Wilczek's work could make you wonder whether the Higgs really deserves credit for the origin of mass. It does, but there is a subtlety to its role that is often missed in popular descriptions. The Higgs field is responsible for making quarks, electrons, and other particles heavy in the first place. Without their masses—minuscule though they are—these particles would never form atoms like the ones we know. Only once quarks have come together does the final, much greater mass of the protons and neutrons they make up emerge.

Wilczek helped to overturn a popular media myth about the Higgs particle by clarifying that it is not directly responsible for all of the mass in the universe. Instead, it accounts for a tiny but crucial first endowment of the mass we know is out there.

Alan Guth works down the hall from Wilczek in the physics department at MIT. Whereas Wilczek's calculations took him deep inside the atom, Guth's transported him back in time to the creation of the cosmos. There, he found evidence to suggest that in the beginning, something very much like the Higgs field did a lot more than imbue particles with mass.

The story of Guth's breakthrough begins at Cornell University in 1978. At the time, so-called Grand Unified Theories (GUTs) were shaping up to be the next big thing. GUTs aimed to unify all of the known forces in nature—except for gravity. That meant finding a single theory that, in one fell swoop, described electromagnetism, the weak force, and the strong force at work inside atomic nuclei.

One day, a friend of Guth's at Cornell asked for help. Henry Tye wanted to know if GUTs predicted the existence of a hypothetical particle that Paul Dirac had toyed around with in the 1930s. The particle was a strange one, to say the least: it was effectively a microscopic magnet with only one pole. Normal magnets always have two poles, one north and one south. The quirky nature of the mysterious particle led to an obvious name: scientists called it the "magnetic monopole."

Guth's calculations showed that GUTs did predict the existence of magnetic monopoles, but there was a problem. "I told Henry he might as well forget about them. No one was ever going to make one in a particle accelerator, because they were ridiculously heavy. To me, they were just another untestable prediction of GUTs," Guth recalls. By ridiculously heavy, he means something like the mass of an amoeba.

Tye wasn't happy. He came back and suggested the two of them work out how many of these strange magnetic particles were created in the Big Bang. "Now that was a really crazy idea," says Guth. He kept his mind on other things, while Tye kept dropping by in the hope of persuading him to help look into it.

Guth steered clear of Tye's idea until a lecture at Cornell changed his mind completely. The physics department regularly invited scientists from other universities to come and speak, and this time it was Steven Weinberg's turn. Weinberg, who was just a year away from winning the Nobel Prize, had become one of the leading proponents of GUTs. The depth and clarity of his talk convinced Guth that the theories might not be so crazy after all.

Right after Weinberg's lecture, Guth talked to Tye, and they agreed, finally, to work out how many magnetic monopoles were made in the Big Bang. Their results were mystifying. They showed that the universe should be chock-full of the strange particles. If that was the case, the universe would be hundreds of trillions of times more massive than cosmologists knew it to be. That wasn't the only problem. Guth calculated that the extra mass of all these particles would have caused the expansion rate of the universe to slow to its present value in just 30,000 years. "I got all kinds of comments," said Guth.

Guth's calculations showed that if the Grand Unified Theories scientists were talking about were correct, several Higgs-like fields would have existed and tied themselves in knots in the first fleeting moments after the Big Bang. These knots would have formed the magnetic monopoles he was trying to count. Guth realized that something must have happened in the early universe to prevent this from happening, though. The answer, he realized, lay in a phenomenon called "supercooling."

An example of supercooling is when water is chilled to below its freezing point of zero degrees Celsius without forming ice. This supercooled water is in a "metastable state"—drop an ice crystal into it and the liquid immediately freezes solid. Guth reasoned that the universe might have supercooled as it expanded. Under these conditions, the Higgs-like fields would have become less tangled, creating far fewer magnetic monopoles.

Guth's real breakthrough followed directly from this train of thought. One evening, he returned to his apartment and worked out what a period of supercooling might have meant for the expansion of the early universe. "There was a step in the calculation when I suddenly saw what was happening. The expansion rate was exponential. It was unbelievably exciting," Guth says.

What Guth had discovered has become known as "cosmic inflation." It says that, for a fraction of a second after the Big Bang, the universe suddenly expanded at a breakneck pace before settling down to a more sedate rate of expansion.

Cosmic inflation theory is a work in progress and scientists' views on it are not set in stone. Guth's original theory has been revised

more than once since he first described it in 1981. But most physicists are generally agreed on one thing: the energy driving cosmic
inflation came from a field that behaves very much like the Higgs
field.[8] Some think it was the Higgs field as originally described in
1964. The field has its own particle, the aptly named "inflaton."

Every winter, the LEP collider at CERN was switched off for a
few months as part of a deal with the local energy provider that
ensured it wasn't running when electricity demand from the surrounding communities reached its annual peak. The downtime was
used by engineers to carry out maintenance and necessary repairs.

On the morning of February 13, 1995, staff arrived to find an unbelievable scene. The main control room of a smaller accelerator that
was used to fire particles into LEP had been practically dismantled.
Cabling had been pulled from the equipment and around 1,200 electronic modules that were needed to control the accelerator were gone.
Only days before LEP was due to switch on again, the equipment
had been sabotaged.

The culprit was later tracked down. Nicolas Blazianu, a well-
regarded control room operator who had worked at CERN for
twenty-seven years, had spent the entire weekend removing and hiding the components. He then disappeared, but telephoned the CERN
management to say he would tell them where the parts were for 2 million Swiss francs.

Blazianu eventually gave himself up to police in Bourg-en-Bresse
in southern France and revealed that he had meticulously hidden the
components around the CERN site: in the ceiling, under the floors,
and behind walls. Staff believed that Blazianu had gone beserk after
a tiff with his ex-wife, who still worked at CERN as an administra-
tive assistant.[9]

The bizarre crime was a serious blow to CERN: the electronics
used to run the sabotaged accelerator had been developed over twenty
years, and documents showing where the different components

should go were incomplete. Even if the parts had survived the ordeal, the machine would need to be tested extensively after everything was put back together.

Blazianu was charged with theft and attempted extortion. CERN reviewed its security procedures and got its engineers working flat out to rebuild the damaged control room. Somehow, they managed to get the machine fixed without incurring a major delay.

The following winter, CERN was hit by another case of sabotage. The machine was down for longer than usual so that contract engineers could upgrade it to run at a higher energy. When it came to switching LEP back on, in June 1996, neither of the two particle beams in the machine would circulate properly. At first, the problem was put down to yet more teething troubles, but after five days of checks, engineers figured out that there was some kind of blockage in a 10-meter stretch of the 27-kilometer ring.

When the collider ring was opened up, the CERN engineers peered down the beam pipe. In each direction, they saw what looked like a green glass lens lodged in the tube. They looked for a pole to poke them out with and eventually retrieved the offending objects. They were two Heineken beer bottles, their labels scorched by the particle beams that had been colliding into them.

Though CERN officials made jokes about the beer's advertising slogan—"the beer that reaches parts other beers can't reach"—it was a serious act of vandalism, and the police were called in to take fingerprints from the bottles.

"It was the kind of malicious act you always worry about," Evans says. "People do awful, mindless, idiotic things, especially when the machine is being closed up. It's a reminder that ill intent is never very far away."[10]

The LEP collider's main role was to churn out W and Z particles, which it did by the million, so scientists could study them in depth. While this was going on, scientists working on the machine's four large detectors, Aleph, Opal, L3, and Delphi, scoured their data for telltale signs of the Higgs boson. Somewhere amid a decade's worth of collisions, they hoped to see debris patterns that were the unmistakable signatures of the long-sought-after particle.

CERN was committed to running LEP until 2000, after which it was to be closed down, ripped out of the ground, and replaced with the far more powerful Large Hadron Collider. As the deadline approached, technicians pushed LEP harder and harder, and by the spring of 2000, both beams were charging around the collider's ring with energies far higher than the machine was designed to handle. Still there was no sign of the elusive Higgs boson.

Shutting LEP down would mean calling off the hunt for the Higgs particle at CERN for at least five years while the LHC was built. During that time, Fermilab's Tevatron would be the only collider with a realistic shot at finding the particle. The Tevatron had been switched off in 1996 to upgrade its detectors and improve the machine's performance. The U.S. collider was on target to rejoin the hunt for the Higgs particle in 2001, a year after LEP was due to be switched off. The deadline put CERN under pressure to drive the LEP collider far beyond its design capability, in case the Higgs was just within its reach.

A few months before CERN planned to pull the plug on LEP for good, one of the collider's detectors flickered in a way it had never done before. One of the other detectors twitched, too. The scientists on the machine felt their hearts skip a beat. "We'd squeezed LEP to its limit," Evans recalls. "We were just getting ready to switch it off. And then all hell broke loose."

8

The End Is Not Nigh

Somewhere below street level on a highly populated island directly east of Manhattan, an unfortunate incident is about to happen. A particle collider that has been running for years without a glitch is crashing gold ions together, as it has done since it was first switched on. There is so much energy in the collisions that protons and neutrons inside the ions melt on impact, releasing a hot soup of quarks and gluons.

The next collision that takes place is different from any that have gone before. Usually, when the ions collide, the quarks that are released huddle back together and form harmless subatomic particles. This time, they recombine in a way scientists considered so unlikely they ruled out the possibility of it ever taking place. The unusual speck of matter is flung out of the main accelerator chamber and gets lodged in a giant magnet surrounding one of the detectors.

Once embedded, the fragment behaves in a curious manner. It begins to attract and engulf the atoms around it. As it swells in size, it brushes up against more and more of its atomic neighbors and consumes those too. When it has grown to the size of an ordinary atom, it falls unnoticed to the laboratory floor, where it promptly sinks through the concrete and into the ground.

Out of sight, the growing shard spirals down toward the center of our spinning planet, transforming matter as it goes and releasing enough heat to melt rock and ore. Before long, the ground beneath southeastern New York City starts to shudder and creak in a terrible process that pulls in more and more of our world. Later, after cities have folded and oceans boiled away, the planet crunches in on itself. All that is left of Earth is a quiet, hot ball of matter that could sit inside PNC Park in Pittsburgh without touching the sides.[1]

Frank Wilczek didn't usually spend his summers pondering such crazy doomsday scenarios. His normal routine was to head for New Hampshire and enjoy the sun and change of pace at his out-of-town retreat. There was no phone up at the house there, so anyone who wanted him had to wait. That was how it usually worked, anyway.

The summer of 1999 was different. Weeks earlier, a couple of letters had arrived at the offices of *Scientific American* magazine. They had raised concerns about a new particle collider that was being commissioned on Long Island. The magazine had published a piece about the machine, the Relativistic Heavy Ion Collider (RHIC), known informally as "Rick," a few months earlier under the title "A Little Big Bang."[2] Rick was designed to crash gold ions together so scientists could study an exotic type of matter called "quark-gluon plasma" that was thought to exist during the earliest moments of creation.

One of the letters about Rick came from Michael Cogill in British Columbia. "I am concerned that physicists are boldly going where it may be unsafe to go," he wrote. "What if they somehow alter the underlying nature of things such that it cannot be restored?" The other letter came by email from Walter Wagner in Hawaii. He asked whether scientists were absolutely sure there was no danger of Rick unwittingly creating a black hole that could swallow the Earth in seconds.

The letters were the starting point for the particle physics equivalent of a media-driven health scare. The readers' queries were legitimate in that they raised interesting points of science, but the events they set in train were almost as ludicrous as the doomsday scenario described earlier. The debacle culminated with legal action that threatened to close American and European particle colliders, an outcome that would have crippled particle physics and brought the hunt for the Higgs boson to a halt.

Though the scare stories centered on the Rick machine at Brookhaven, they came at a sensitive time for both CERN and Fermilab. The European laboratory planned to close its LEP collider the following year and replace it with the more powerful Large Hadron Collider. At Fermilab, engineers were close to finishing a five-year project to boost the Tevatron's performance. The detectors had been given a $200 million makeover that enabled them to handle 10 million collisions per second. And with a new main injector, the beams of particles circulating in the Tevatron would be twenty times more intense when the machine switched back on. It was not a good time for a public scare over particle colliders.

The editors at *Scientific American* decided to get a response to the readers' queries from a leading scientist in the field. They called Wilczek and asked if he would oblige. He wrote out a reply and sent it to the magazine shortly before leaving for New Hampshire. The piece would appear alongside the letters in the July issue of the magazine.

Wilczek explained that black holes could hardly be made at Rick, but his reply didn't end there. The piece went on to highlight another deeply speculative but "quite respectable possibility" that a new stable form of matter called a "strangelet" might be created in the collider. "One might be concerned about an 'ice-9' type transition, wherein a strangelet grows by incorporating and transforming the ordinary matter in its surroundings," he wrote. His response ended on the reassuring note that even a rogue strangelet was an implausible agent of global destruction.

The reply seemed harmless enough. It reassured readers they needn't fear black holes beneath New York any time soon, and introduced an intriguing theoretical concept as brain food. "I thought I'd use the opportunity as a teaching experience," Wilczek recalls. "It was scientifically interesting to mention strangelets, so to make my reply more appealing, I said something along the lines of, well, if you really want to worry about something, worry about this."

When Wilczek's reply reached *Scientific American*, the editors realized it was too long and cut it down by around a third.[3] The edits changed the tone of the piece significantly. In edited form, Wilczek's response seemed more equivocal about the dangers of marauding strangelets. "They made it much less clear that the strangelet scenario was also highly implausible," Wilczek said.

At Brookhaven National Laboratory, where the collider was being readied for action, the director, John Marburger, heard the magazine was about to publish the letters and Wilczek's response. Marburger had been parachuted in to take charge of the government facility a year earlier, when a harmless leak of radioactive material from a research reactor on the site had sparked a public outcry.[4] The backlash against the laboratory was so intense that the Department of Energy shut the reactor down and ordered a multimillion-dollar cleanup operation.

Marburger, who later served as science adviser to President George W. Bush's administration, saw a storm on the horizon. *Scientific American* was a serious magazine. Wilczek was a renowned and brilliant physicist. The local community was distrustful of the government and its running of the lab. The letters, and especially Wilczek's response, had all the ingredients of a public relations disaster, if nothing more globally catastrophic. "I realized immediately this would be a problem," Marburger recalled. "Even a small probability is a big deal when we're talking about destroying the Earth."

Before *Scientific American* went to press, Marburger assembled a committee of physicists. He asked them to look into a range of far-fetched disaster scenarios that the collider might even remotely risk setting in motion. "The first person I contacted to be on the committee was Frank," Marburger said. "My attitude was, all right Frank, you started this mess, now you have to help to fix it."

Within days of *Scientific American* hitting the magazine racks, Marburger's storm arrived. On July 18, the *Sunday Times* in London ran a story under the headline "Big Bang Machine Could Destroy Earth." It talked of Rick being "under investigation" and claimed Marburger had commissioned experts to report back on "whether the project could go disastrously wrong." An accompanying editorial lamented: "The men in white coats would send us, and them, into the oblivion of a black hole of their making." British newspapers are not renowned for their restraint in the summer months.

Other media inevitably picked up on the story. Some called Rick the "doomsday machine." The press office at Brookhaven was over-run with calls. One person wanted to know if a black hole created at the machine had downed John F. Kennedy, Jr.'s, plane.[5] In a state-ment designed to diffuse the situation, Marburger reminded people that "scientists are no more willing to endanger the world, or them-selves, than anyone else."

Everyone wanted to talk to Frank Wilczek, but he had already left for New Hampshire. "I don't have a telephone line there and at that time we didn't have cellphones, but I was besieged by press from all around the world. I had to drive up the road where there was a payphone and answer them," Wilczek said.

The furor over the collider wasn't entirely negative. The *Sunday Times* article guaranteed publicity for the machine around the globe, and for every news article that proclaimed the end of the world, there were others that remained more level headed and explained the value of the Brookhaven experiments.

In Geneva, the CERN management realized the European lab was an obvious target for a vocal minority who made it clear they wanted particle colliders closed down. The giant LEP machine had only a year or so left to run and hadn't destroyed the world yet. What unnerved CERN more was the prospect of a public backlash against the Large Hadron Collider, which was destined to become the most powerful particle collider in the world. If the machine was scrapped because of a collapse in public support, the Higgs particle might never be found. A range of other theories would also be kicked into the hinterland of untested ideas. As a precaution, CERN's director general, Luciano Maiani, appointed his own team of physicists to review the safety of the new collider.

The investigations at Brookhaven National Laboratory and CERN surely marked the first time governments had called on scientists to answer questions about whether laboratory experiments might risk destroying the planet. The only comparable situation was in 1945, when the Manhattan Project scientists Emil Konopinski and Edward Teller had calculated the chances of nuclear bombs igniting the Earth's atmosphere. They decided it was impossible, at least with the bombs they had to hand, though that didn't stop Enrico Fermi from taking bets on whether the world would end when the first atom bomb was detonated on July 16 that year at the Trinity test site.

The safety committees put together by Brookhaven National Laboratory and CERN looked at scenarios that had been dreamed up as potentially catastrophic for the planet. They included the accidental creation of dangerous strangelets (some varieties are thought to be entirely harmless); the production of a black hole that consumes the Earth in the blink of an eye; the creation of atom-destroying magnetic monopoles; and a profoundly terminal scenario known innocuously as "vacuum decay."

Wilczek's reference to "ice-9" was drawn from Kurt Vonnegut's 1963 novel *Cat's Cradle*, in which he describes a world transformed by the accidental release of an alternative and more stable form of ice developed by the military. Ice-9 was designed to banish mud, and with it the prospect of the army's troops and vehicles getting bogged down in the stuff. A single chip of ice-9 tossed into a muddy expanse becomes a "nucleation site," a point where surrounding water molecules latch on and freeze into a new crystalline structure as hard as a desk. Unlike ordinary frozen water, ice-9 doesn't melt until it's warmed up to 45.8 degrees Celsius.

A year before *Cat's Cradle* was published, scientists in Russia had created something that had raised fears similar to the fictional ones. Nikolai Fedyakin was a chemist working in an obscure government laboratory in Kostroma, a city in central Russia. He spent his time studying the behavior of water in thin glass capillary tubes. One day, he was examining some he had half-filled with water. Above the original water level, he noticed that new, separate specks of water had appeared. Over time, they grew at the expense of the water beneath. Tests showed that these new water droplets were far denser than ordinary water. Fedyakin was stunned. His experiments pointed to a new phase of water, one that was more stable than regular water and capable of transforming regular water on contact.

Major discoveries don't stay in minor laboratories for long. When word of the breakthrough got out, Fedyakin's work was transferred to Moscow, where one of Russia's preeminent scientists, Boris Deryagin, took over. Deryagin, who was renowned as a meticulous and thorough scientist, repeated Fedyakin's experiments successfully. He declared the discovery of "anomalous water," a new and previously unknown phase of life's most precious liquid.[6]

The broader scientific community was deeply skeptical until a team of scientists from the U.S. Bureau of Standards published a study that supported the Russian findings. They looked at how anomalous water absorbed infrared light and claimed it was different from ordinary water. They even went so far as to rename the liquid "polywater," because they were convinced the water molecules had come together to form a polymer-like gel, made up of long chains and hexagonal rings of water.

Polywater appeared to share some disturbing properties with ice-9. Some scientists thought it was more stable than ordinary water and melted at a higher temperature than ice. If that was true, and if it was ever manufactured and released into the world's watercourses, it might polymerize the planet's water supply. In other words, water molecules would link together to form giant molecules like those found in plastics. The consequences for life on Earth were too terrible to contemplate. In October 1969, the journal *Nature* published a letter from a Dr. F. J. Donohoe at Wilkes College in Pennsylvania, who demanded assurance from scientists that polywater was safe.

"The consequences of being wrong about this matter are so serious that only positive evidence that there is no danger is acceptable," Donahoe wrote. "I regard the polymer as the most dangerous material on Earth. . . . Scientists everywhere must be alerted to the need for extreme caution in the disposal of polywater. Treat it as the most deadly virus until its safety is established."

Polywater gained credibility, but plenty of scientists were still unconvinced. They argued that if there really was a more stable form of water, then we would surely have found gloopy films of it long ago. The physicist Richard Feynman pointed out that, if polywater was real, millions of years of evolution would have given us a creature whose sole means of survival was drinking water and excreting polywater, because the transformation released energy the organism could thrive on.

As Feynman suspected, polywater turned out to be nonsense. After years of experiments, scientists determined that the only thing truly distinguishing polywater from normal water was that it contained impurities, many, it seemed, from the glass capillaries it was kept in. From excited discovery to embarrassed dismissal, the diversion that polywater created spanned a decade.

Fears of polywater congealing the world's rivers and oceans quickly subsided, but the prospect of exotic Earth-transforming matter did not. A few years after the fuss died down, Chinese American Nobel Prize–winning physicist Tsung-Dao Lee and Italian theorist Gian Carlo Wick speculated that particle colliders might be able to slam atomic nuclei together so hard that they collapsed into a stable and incredibly dense form of matter. Lee was so excited by the idea that he proposed joining two particle accelerators together to see if scientists could make some.

Engineers at Lawrence Berkeley Laboratory in California got to work. They rigged two accelerators together so that one fired atomic nuclei into the other, which accelerated them even more and slammed them into a target. When the machine was switched on in the mid-1970s, scientists weren't sure if they would make Lee and

Wick's "abnormal matter." What they did know was that, if they did, it might not be entirely safe.

In May 1979, years after the "Bevalac" switched on, but before it started accelerating heavy ions like uranium, senior scientists met in secret at the laboratory to discuss whether the Bevalac was likely to create abnormal matter, and if so, whether it posed a risk.[7] Once again the ice-9 scenario loomed large. If abnormal matter was more stable than ordinary matter, a tiny amount might conceivably trigger a global disaster by converting all the matter it came into contact with.

The assembled experts, among them Tsung-Dao Lee and Bernard Harvey, the associate director of nuclear physics at Lawrence Berkeley Lab, spent a day and a half talking through the disaster scenario and whether the experiments at the Bevalac should be aborted. When the meeting came to a close, they were unanimous in believing that abnormal matter wasn't a danger. More energetic collisions had been happening for billions of years on the moon when particles in cosmic rays hurtled into the lunar surface. If abnormal matter was dangerous, it would have destroyed the moon long ago. Since the moon seemed to be doing fine, the scientists concluded there was no reason to worry.

For researchers involved with the machine there was another, more personal risk to consider, though most undoubtedly considered themselves safe. From the late 1970s to the mid-1990s, the United States had endured an erratic run of bombings that had targeted scientists and airline workers. FBI agents working on the case knew that their quarry, the Unabomber, was deeply opposed to what he perceived to be the blind march of technology and its potential consequences for humanity.

A year before the Bevalac was switched off, two physicists, Gary Westfall at Michigan State University and Sabul Das Gupta at McGill University in Montreal, wrote an article for the magazine *Physics Today* that celebrated the Bevalac's achievements. It mentioned that "meetings were held behind closed doors" to decide whether the risk of a catastrophe was sufficiently serious that the experiments should be scrapped on grounds of safety. The piece went on to add: "Experiments were eventually performed, and fortunately no such disaster has yet occurred."

When the article appeared, the FBI feared that Westfall and Das Gupta might become targets for the Unabomber. Both were put on a watch list, and Westfall agreed to have his mail searched for explosives. Das Gupta refused, instead choosing to place his trust in the Canadian postal service and his university secretary. No explosives were ever intercepted on their way to the researchers.

By the time the Rick collider at Brookhaven was ready to be switched on, concerns about Lee and Wick's abnormal matter had faded. But Wilczek's piece in *Scientific American* ensured that insatiable marauding strangelets were on hand to take over the role of scary world-ending entity. Scientists came up with the idea of strangelets while pondering what might happen if the protons and neutrons inside atomic nuclei were squeezed to extraordinary pressures. Such a thing might happen naturally in the heart of neutron stars, which form when ordinary stars explode and collapse under their own gravity. Neutron stars are staggeringly dense objects: a teaspoon of material from the core of one could weigh around 100 million metric tons.

Normal protons and neutrons are made up of two kinds of quarks, known as up and down, but scientists suspect that some of these, under great pressure, might turn into a third variety, known as "strange quarks." The resulting mix of quarks is a strangelet. In 1984, Ed Witten, a physicist at the Institute for Advanced Study in Princeton, whom many regard as Einstein's natural successor, calculated that, once created, strangelets might hang around even once the enormous pressure needed to make them has been released. The paper seeded the suspicion that if strangelets were more stable than ordinary matter they might trigger the kind of ice-9 scenario Wilczek described.

The safety panels at Brookhaven and CERN came up with lengthy theoretical explanations of why there was no reason to worry about strangelets at either collider.[8] If they existed at all, they were hard to make. If they could be made, they weren't very stable. And if they did happen to linger around for longer than expected, they were almost certainly positively charged, in which case they wouldn't be able to draw in and eat atomic nuclei.

Theoretical assurances about the planet's safety only went so far. The Harvard physicists Sheldon Glashow and Richard Wilson summed up the feeling of uneasiness in an article in *Nature* in December 1999:

> If strangelets exist (which is conceivable), and if they form reasonably stable lumps (which is unlikely), and if they are negatively charged (although the theory strongly favours positive charges), and if tiny strangelets can be created at Rick (which is exceedingly unlikely), then there just might be a problem. A newborn strangelet could engulf atomic nuclei, growing relentlessly and ultimately consuming the Earth. The word "unlikely," however many times it is repeated, just isn't enough to assuage our fears of this total disaster.

To strengthen their conclusions, the safety panels argued that Mother Nature had already done the Rick experiments for us. Cosmic rays contain metal ions that move at nearly the speed of light. These crash into minerals on the moon and on asteroids and into free-floating ions in clouds of interstellar dust and gas. If dangerous strangelets were easy to make, they would have been made in space by now.

As with the Bevalac and concerns over abnormal matter, the enduring presence of the moon was cited as a strong indication that malicious strangelets were at least extraordinarily rare. If 5 billion years of being battered by cosmic rays hadn't turned the moon into a giant lump of strange matter, five years of collisions at Rick would hardly be a threat to Earth. Further reassurance came from considering the fate of asteroids. If cosmic rays had created "killer asteroids" by turning them into strangelets, some would inevitably have fallen into the sun or other stars and destroyed them. But as scientists looked around at the 70 billion trillion stars in the visible universe, they saw nothing more than the natural die off of stars through supernova explosions.

To get a feel for just how unlikely the Large Hadron Collider was to turn up a nasty surprise, CERN scientists did a calculation. Our own sun is struck constantly by cosmic rays with at least as much energy as the LHC collisions. Based on the number of stars in the observable universe, they estimated that nature has conducted the entire lifetime (around twenty years) of LHC experiments 10 million trillion trillion times over since the universe began. What's more, the impact of cosmic rays on distant stars collectively perform the LHC experiments again 10 trillion times over every second.

Of all the disaster scenarios the scientists at Brookhaven and CERN reviewed, the creation of a black hole that swallows the planet whole received the most media attention. Both groups dismissed the danger. To make an ordinary black hole, a particle collider would have to compress a truly staggering number of particles into such a tiny volume that gravity would make the ball of matter collapse in on itself. The feat was so far beyond the capability of any known collider—and any in the foreseeable future—that both teams spent little time on the scenario.

The dismissal assumed that Einstein's equations were the final word on gravity, but that is unlikely to be the case. Some recent theories call for nature to have hidden dimensions that are curled up so small we cannot see them. As yet, there is no evidence to suggest that we live in a world of more than four dimensions—three spatial ones, plus time—but if extra dimensions do exist, microscopic black holes could conceivably be made very easily in modern particle colliders.

Again, scientists reasoned there was nothing to fear even if microscopic black holes were created in their colliders. In 1975, the Cambridge cosmologist Stephen Hawking showed that black holes give off heat. The smaller they are, the more heat they lose. According to theories that incorporate extra dimensions, any black holes created at the LHC would be microscopic runts, around a millionth of a billionth of a millimeter wide. At that size, they would be pinpricks of extreme heat—around a billion times hotter than the center of the sun. The good news is that they would lose heat so fast they would evaporate out of existence in an instant.

Another potential disaster scenario the physicists found easy to dismiss involved the creation of magnetic monopoles. These are truly bizarre particles, and as Alan Guth found immediately before stumbling upon the idea of cosmic inflation, they are far too heavy to be created in any accelerator that can be envisaged today. The heaviest particle ever made at a particle collider, the top quark, which weighs around 170 GeV, was discovered at the Tevatron in 1995. Magnetic monopoles, if they exist, are likely to be more than a trillion times heavier.

Out of curiosity, the CERN safety panel worked out how much damage a magnetic monopole could cause if one happened to crop up at their machine. Some theories say that magnetic monopoles can be dangerous because they transform protons and neutrons into electrons, positrons, and other particles. Essentially, they vaporize normal matter. The CERN panel found that a magnetic monopole would destroy only half a gram of ordinary matter before the energy it released in doing so propelled it out into space.

There is a finality to the end of the world, however it transpires, that sets it apart as the worst possible future one can imagine. Certainly that is true for humanity and for the millions of species we share the planet with. But in the fourth doomsday scenario, scientists were forced to consider an even worse fate than the catastrophic destruction of our home planet and all that lives on it. The fourth doomsday scenario, vacuum decay, doesn't only extinguish all life on Earth. It destroys the prospect of any hope of life for the rest of time across great swathes of space.

To the scientists of the mid-seventeenth century, a vacuum was what you were left with if you attached one of those newfangled pumps to a glass vessel and set your mind to making the damn thing work. If you were persistent and managed to remove all the air from inside, you had a truly empty vessel: a container full of absolutely nothing.

To modern scientists, the vacuum is far from empty. It holds an unseen world of thriving fields and associated particles that are constantly popping in and out of existence. Energy is locked up in these fields and gives rise to what scientists reasonably refer to as the universe's vacuum energy.

A universe is in its most stable state when it contains as little energy as possible. The thing is, scientists don't know whether our universe is in its most stable state or not. The vacuum might contain more energy than is strictly necessary, in which case, given the right incentive, it might suddenly give way and crash into a more stable, lower-energy configuration.

You can see a similar process at work in your front room. When you proudly balance a photograph of your grinning in-laws on your mantelpiece, it has a certain potential energy. Gravity is just dying to reduce that potential energy by pulling it down into the fireplace below. All it is waiting for is an unfortunate breeze or a gentle nudge to destabilize the picture so it can pull it from its perch.

If the universe is in a similarly precarious state—and it's a big if—a kick of energy could conceivably knock it off its own cosmic perch and down into a more stable state. Scientists at Brookhaven and CERN wondered if the energy created inside a particle collider such as Rick or the LHC might give the universe just such a kick. The consequences, were it to happen, are extraordinary.

The Harvard physicist Sidney Coleman had a fine way with words. It was he who had suggested that his class tear Peter Higgs to shreds over his big idea, the day after Higgs gave his major seminar at the Institute for Advanced Study in Princeton in 1966. Coleman was intrigued by the prospect of humanity living in a universe that was only metastable, that is, functioned perfectly well, but in the right (or wrong) circumstances could collapse into a lower energy state.

Coleman worked out what would happen if, in some corner of the universe, vacuum energy for some reason or other suddenly fell off the mantelpiece—that is, dropped from what appeared to be a stable configuration to a far more stable one. He showed that a bubble of "true vacuum" would spring into existence and grow with astonishing rapidity. The edge between the true vacuum bubble and the "false" vacuum of the old universe would race outward at the speed of light.

The energy in the vacuum is the bedrock that natural laws rest upon. Were it suddenly to drop, the laws of physics would change in an instant. It would be toast for us and every other creature on Earth, but Coleman found something even more disturbing. Our old universe would be replaced by a new and more stable version, but one in which the conditions were such that life would never again gain a foothold.

Ten pages into an eleven-page paper he published in 1980, Coleman summed up his conclusions in what is surely one of the most arresting paragraphs ever to appear in a scientific journal:

> The possibility that we are living in a false vacuum has never been a cheering one to contemplate. Vacuum decay is the ultimate ecological catastrophe; in a new vacuum there are new constants of nature; after vacuum decay, not only is life as we know it impossible, so is chemistry as we know it. However, one could always draw stoic comfort from the possibility that perhaps in the course of time the new vacuum would sustain, if not life as we know it, at least some structures capable of knowing joy. This possibility has now been eliminated.[9]

A couple of years after Coleman's striking paper was published, Wilczek and a colleague, Michael Turner, tried to work out whether our universe was in its lowest energy state. Writing in the journal *Nature*, they concluded: "It seems distinctly possible . . . that our present vacuum is only metastable, and that nevertheless, the universe would have chosen to get 'hung up' in it. If this is the case, then, without warning, a bubble of true vacuum could nucleate somewhere in the universe and move outwards at the speed of light."[10]

Since nothing travels faster than the speed of light, we would have no notice that such a cosmic catastrophe was heading our way. "You don't see anything. You're just gone," says Wilczek. "We'd all become purple haze."

Even if we wouldn't know what hit us, it is hard to suppress a morbid fascination with what would physically happen to us and our surroundings. One possibility is that, as the new vacuum swept by, the strong force that binds particles together inside atomic nuclei would suddenly become short range, just like the W and Z bosons do when the Higgs field switches on. If that happened, the atoms inside everything would simply and spontaneously fall apart. Just like that.

Were it needed, and it surely was, reassurance of a sort arrived the following year in the form of a note from Martin Rees, who later became Britain's astronomer royal and president of the Royal Society, and Piet Hut, a physicist at the Institute for Advanced Study in Princeton. Once again, the antics of cosmic rays were on hand to calm humanity's existential fears.[11]

Hut and Rees pointed out that the universe has lasted perfectly well with its present vacuum for nearly 14 billion years. The implication was that humans would have to do something more violent than has occurred anywhere else in the universe to cause the vacuum to decay.

The most energetic particle collisions on Earth take place when ions in cosmic rays slam into ions in the upper atmosphere. The duo calculated that around 100 million collisions happen in our atmosphere every second that are more violent than any that could be produced in modern particle colliders.

The message from Hut and Rees was that the universe wasn't so fragile that it could be torn apart by an earthly particle collider. Well, not yet anyway. The question would have to be asked again if particle colliders ever became a hundred times more powerful than the ones we have today. For now, we are safe. "We can be confident that no particle accelerator in the foreseeable future will pose any threat to our vacuum," the scientists wrote in *Nature*. The statement was welcome, even if it does conjure up an image of protesters gathering at the gates of CERN in twenty years' time waving placards scrawled with messages like "Hands Off Our Vacuum."

It wouldn't be the first time protesters had turned up at a particle collider with hopes of bringing an end to the end of the world. In the mid-1990s, a small group of protesters, worried about scientists "tearing a hole in the universe," picketed Fermilab when the Tevatron collider was being switched on after a revamp. The protest was a modest affair instigated by Paul Dixon, a psychologist at the University of Hawaii, who made a shoulder-high banner out of a bedsheet declaring Fermilab to be the "home of the next supernova."

Both the Brookhaven and CERN safety groups used the cosmic ray argument Hut and Rees had worked through to confirm that their particle colliders were in no danger of causing vacuum decay. The question will surely arise again when more powerful colliders are built.[12]

The original Brookhaven safety report was published in September 1999. It wasn't written for the public, but it did become a cornerstone of a wider public relations exercise. Its value, to Marburger at least, was that it contained the unanimous—and overwhelmingly positive—conclusions of four scientists who were arguably the best in the world. On the basis of the report, Marburger reassured the public that Rick was not a threat to the planet.

Walter Wagner, whose letter had raised specific fears about black holes, was one of those who wasn't reassured. Before his letter had even been printed in *Scientific American*, he had applied through a California court to have a temporary restraining order put on Brookhaven's Rick. When the application was turned down, Wagner asked the court to reconsider. After three such pleas, the court dismissed the lawsuit completely.

Wagner is an unlikely character.[13] As a former radiation officer in San Francisco, he was known to set off on neighborhood radioactivity hunts. He became concerned about the uranium glazes used in decorative tiles and began testing the ones he found on public buildings as well as in residential areas. He would knock on strangers' doors and, waving his Geiger counter, offer to come in and check their bathroom tiles. At a science conference, Wagner set up a stall to publicize what he regarded as unsafe radiation levels at a local school. The state health department considered his readings so alarmist that they set up their own stall nearby to counter Wagner's message.

In the years following his attempt to close Rick through the courts, Wagner tried similar tactics to prevent CERN's LHC from being switched on.

Wagner was among an extreme few who became prominent in the media for opposing particle colliders, but that is not to say the broader public was completely sold on the safety of the machines. Several polls suggested quiet concern among a substantial portion of society. Some 66 percent of people who took part in an online poll arranged by the BBC prior to the LHC starting up believed the machine was too dangerous to be switched on. Sixty-one percent of the respondents in an AOL news poll conducted at about the same time felt the same way. It is highly unlikely that these polls were accurate reflections of public opinion: the individuals who responded were self-selected and voted after reading news stories that highlighted the proposed dangers of the machines.

The flaws of simple polls conducted by the click of a mouse or a quick telephone call are well appreciated. They take no account of people who decide not to vote because they don't care about the issue. Those who do vote are dominated by the extremely interested, who are not beyond voting multiple times. A now infamous poll run by *USA Today* in 1990 asked readers to vote on whether business magnate Donald Trump was symbolic of what made America a great country. Of more than 6,000 telephoned votes, 81 percent agreed that Trump was a figure of greatness. It later emerged that nearly three-quarters of the votes were called in from just two telephone numbers. Comparisons between such basic polls and more scientific ones show that the results they produce can differ by tens of percentage points.[14] Peter Higgs, for one, thought the whole idea of colliders destroying the world was nonsense.

One problem that continues to plague discussions over the safety of particle colliders—though the issue is relevant to other areas of cutting-edge science as well, such as synthetic biology and genetics—is that it is nearly impossible to have an objective and informed public debate on the issue. The people who understand the concerns best are those who work in the field under debate—particle physics—so the accusation of vested interests cannot be avoided. Ironically, the most high-profile opponents to a new technology are often so badly informed that they are quickly dismissed as crackpots, and rightly so. The result is an illusion of public debate. Ill-informed opponents do a disservice to people with genuine interest and concern by squandering the opportunity for an even-handed discussion of the risks.

The media was guilty of supporting an illusory debate over the safety of particle colliders by pitting serious scientists against poorly informed critics. This was unnecessary. Among serious academics, there were objective critics of the Brookhaven and CERN safety reports. Few of those who raised concerns in the spirit of healthy debate had their voices heard outside of technical papers and specialist journals.

John Marburger was well aware that some of the arguments in the Brookhaven safety report had drawbacks that left people open to fearing the worst. For example, the fact that the sun and moon were still with us despite being battered for billions of years by cosmic rays counted for nothing if the catastrophic process people had in mind was spectacularly rare in nature. It might just be pure luck that it hadn't happened yet.

Another cause of confusion was the fact that the risk figure Brookhaven came up with was an upper limit, not a precise probability. The Brookhaven report used cosmic-ray arguments to claim that the chance of a strangelet being created at Rick was less than 1 in 500 million every year the machine was running. Since the machine was expected to work for ten years, this was taken to be equivalent to a 1 in 50 million chance that the enterprise would destroy the world.

The risk figure arrived at in the Brookhaven report was of little real help, then. People tended to interpret it as the odds of the machine destroying the world, but the figure meant something subtly but significantly different. It meant that, if the collisions at Rick were like those in nature, the chances of the machine sparking an unpleasant Armageddon *could not be more than* 1 in 50 million. The true risk of such a catastrophe might be trillions and trillions of times smaller.

The dearth of evidence for or against the existence of strangelets made things even more difficult. To the best of anyone's knowledge, a strangelet has never been made at a particle collider, despite scientists scouring their data for them. They have never been spotted hurtling through space, or sitting around on the moon or on another planet. There is no rigorous theory that says they must exist. Yet no theory specifically rules them out. In a situation like this, calculating a meaningful risk of strangelets being created in a particle collider and then destroying the planet is impossible.

Marburger likens the problem to estimating the risk of being eaten by an elusive Scottish monster when taking a dip in Loch Ness. There is no credible evidence of the monster, and everything we know about science leads us to suspect it doesn't exist; still, no law of nature rules out the monster's possible existence. Actually, the chances of someone being eaten by Nessie are probably much better than the chances of a strangelet disaster occurring. Despite the unlikelihood of Nessie existing, at least there are purported eyewitness accounts from people who claim to have seen the monster. Scientists have no evidence at all for strangelets. In the case of Nessie, believers are undeterred, despite the lack of evidence (and to the relief of the Highlands of Scotland Tourist Board). Strangelet believers have far less to go on.

There was also a question mark over the safety panel's assumption that cosmic ray collisions are equivalent to those in particle colliders. Although nature carries out its own particle collisions by hurling cosmic rays into planets and clouds of interstellar dust, they are not the same as the collisions scientists study on Earth. The differences between the two circumstances may be unimportant, but that is hard to judge when the theories surrounding the particles that might be created are so sketchy.

When ions in cosmic rays smash into the moon, they are traveling at close to the speed of light. If a strangelet was created in the aftermath of a collision, it would have to survive a high-velocity journey through lunar dirt before it had a chance to do any harm. In a collider, the particles meet head on in a vacuum, so the debris created on impact is moving more slowly, and it's not battered by moon rock from the moment it is born. Might a strangelet created under those conditions be more dangerous than one made on the moon?

At Cambridge University, Adrian Kent, a quantum theorist, directly challenged the safety assumptions that had been put forward for the colliders.[15] The Brookhaven report maintained that cosmic-ray arguments alone were enough to guarantee Rick's safety. When Kent pointed out that this conclusion was "gravely flawed," the Brookhaven panel revised its report and removed the claim.

Kent also said that while an upper risk limit of 1 in 50 million might sound low, it has to be balanced against what we stand to lose. If things went wrong, it would be curtains for the world's population of 6.7 billion people, the entire future of humanity, and every other species on Earth.

Kent went on to draw up a calculation that showed that a 1 in 50 million chance of killing 6.7 billion people was *at least as bad* as the certainty of killing 134 people, a figure reached by multiplying the fractional risk by the number of lives at stake. The figure served a purpose—to show that the 1 in 50 million figure wasn't quite as reassuring as Brookhaven scientists might have hoped.

Yet the Rick experiments went ahead, partly on the strength of Brookhaven's initial safety analysis. "As far as I am aware, no effort was made by Brookhaven to reobtain authorization on the basis of [the] revised assessment, or to bring what is a significantly revised case to the media and public attention," Kent commented. "In my opinion, such efforts should have been made."

In the spirit of devil's advocate, Kent pressed on to find more flaws in the arguments scientists had used to justify the safety of particle colliders. On one occasion, he identified a potential disaster scenario none of them had considered. Both safety panels argued that positively charged strangelets were safe because they would repel atomic nuclei around them rather than consuming them. But what if one managed, by some deeply contrived means, to find its way to the sun? Once inside, it could possibly kick-start a catastrophic scenario that destroyed the sun.

Kent argues that detector components from particle colliders could, without anyone knowing, become contaminated with positive strangelets. They might then end up being recycled into spacecraft that one day could be flung out to survey and ultimately fall into the sun. Alternatively, technophile terrorists might obtain some contaminated material and threaten to lob it into the sun, Kent suggests. He warned that "laboratory material potentially contaminated by positively charged strangelets would thus pose a potential danger which, though small, would need careful handling." He concedes that any group with the money and know-how to pull off such a job "could credibly threaten catastrophe by less exotic means, if so inclined."

If the high-energy particle collisions that happen in nature are similar to those planned in terrestrial particle colliders, then theoretical arguments become more valuable. But even here, scientists outside the safety-assessment groups have spotted pitfalls. A risk assessment based on theory has to take into account the chance that the theory in question might be wrong. If the uncertainty in a theory is of any appreciable size, any risk assessment based on it can become worthless. For example, the threat from black holes at the LHC is judged to be minimal. One reason is that, to make black holes in modern colliders, gravity must behave strangely at microscopic scales, something that is postulated but unknown. If they do behave strangely at those scales, and black holes are produced, their harmlessness rests on the correctness of Stephen Hawking's theory, which suggests that they would radiate heat and evaporate into thin air. It is not a cast-iron theory, though, and the details are certainly not nailed down.

The prospect that black holes might, just might, become a nuisance at a particle collider prompted one scientist to dream up an imaginative solution. If the black hole is quite pathetic—as planet-swallowing entities go—it is likely to grow extremely slowly. That would give researchers time to turn a cathode ray tube on the blighter and pump it full of electrons. By swallowing the particles, the black hole would develop a negative charge. Scientists could then trap the black hole in a box lined with negatively charged metal walls. The walls of the box would repel the black hole, so, provided it was in a vacuum, the menacing hole would hover gently while scientists worked out what to do with it. One option would be to load it onto a rocket and blast it out into space. Scientists behind the idea admit that it may have flaws.

Francesco Calogero is an Italian theoretical physicist who spent eight years as secretary general of the Pugwash Conferences on Science and World Affairs. The conferences are a haven for scholars and public servants from different nations to explore ways of resolving conflicts and stop the escalation of arms with frankness and in privacy. In 1995, Calogero accepted the Nobel Peace Prize, which was awarded jointly that year to Pugwash and Joseph Rotblat, a Polish-born physicist who had become a leading proponent of nuclear disarmament. When the fuss broke over Rick, Calogero proposed a better way of assessing its safety.[16]

In 2000, Calogero published a scientific paper that could win awards for its understated title: "Might a Laboratory Experiment Destroy Planet Earth?" In it Calogero argued that Brookhaven should have employed two groups of scientists: a blue team, charged with producing an objective safety report on the collider, and a red team, whose role was to play devil's advocate and deliberately try to prove that the collider experiment was dangerous. When both teams were finished, they should have tried to reach a consensus, and, if possible, put numbers on the agreed risks of going forward.

At the very least, the strategy would have reassured the public that they weren't being hoodwinked by a PR exercise. Setting two teams against each other would have another powerful effect: it would legitimize objective and open criticism, something Calogero found little appetite for among his colleagues. Having solicited expert views on the safety of the Brookhaven collider, he summed up his colleagues' responses as follows: "Many, indeed most, of them seem to me to be more concerned with the public relations impact of what they say, or others say and write, than in making sure that the facts are presented with complete scientific objectivity." Their acquiescence to the official line verged on complicity.

The shortcomings of the risk assessment of Rick were highlighted by Richard Posner, a respected U.S. judge, in his 2004 book *Catastrophe: Risk and Response*. In it, he raises questions over the impartiality of the scientists brought in to judge the safety of experiments and asks what benefit society must expect in order to accept a certain level of risk. Posner called on lawyers to become more scientifically literate, though expecting many to fully grasp quantum theories of hypothetical particles and the intricacies of risk assessment seems ambitious. Ultimately, Posner believes that decisions of this kind should be placed by law in the hands of a permanent "catastrophe-risks assessment board" that could red-light projects if they carried an "undue risk to human survival."

What lessons should we learn from this? History suggests there will always be some world-ending entity lurking among scientists' theories, and the chances of unleashing it by accident will almost certainly be shrouded in uncertainty. If dangerous strangelets and magnetic monopoles are ever ruled out, another possibility will emerge from physicists' theories. How then should society decide whether an experiment that has a minute risk of causing total disaster should be carried out? In the distant past, the consequences of an experiment going wrong affected only those involved or nearby. One argument says that, since particle colliders are primarily of direct benefit only to pure science, we have already come too far. But that is short-sighted. High-energy physics experiments have brought us revolutionary technologies as disparate as the World Wide Web and ion beams for cancer treatment. When we make progress in pure science, technological benefits often follow. Perhaps the best we can hope for is a truly open and public debate in which real risks are laid out. Without that, society as a whole has no chance of making an informed decision. How we achieve this will only become a more pressing issue as science advances.

When Emil Konopinski and Edward Teller sat down to calculate whether an atomic explosion might set off a runaway fusion reaction in the atmosphere, they had enough knowledge of the nuclear properties of the atmosphere to be confident there wasn't any danger. Their conclusions were comforting, but they also marked a point in history where scientists had to take seriously the possibility that the power to destroy the world was theirs to wield. That moment was on Marburger's mind when the exchange of letters in *Scientific American* sparked a flurry of doomsday fears over particle colliders. "The analogy with the Trinity test concerns was on everyone's mind," he recalls. "I wish it had been on Frank's mind when he wrote the response." The Trinity test, on July 16, 1945, in the New Mexico desert, was the first demonstration of a nuclear weapon.

The doomsday scenarios that scientists at Brookhaven and CERN considered seem outlandish. Destroying the planet in spectacular fashion isn't a trivial thing to achieve, and doing so with a particle collider, a machine designed for a quite different purpose, is surely overwhelmingly unlikely. This is the comfort of complacency though: it is easy to think that because we haven't destroyed the world yet, we are not about to do so.

Our planet and the surrounding cosmos may be made of such sturdy stuff that human experiments will never pose a realistic threat. But many physicists are nonetheless mindful of the responsibility that comes with eliciting the workings of nature and harnessing its power. According to Frank Wilczek, things changed when physicists began to understand the realm of quantum physics. He said:

> Classical physics was marvelous in its way, but it wasn't surprising in the way modern physics is. In the quantum world things are different. There are enormous amounts of energy locked up in the small-scale structure of the world, and no one came close to guessing that from everyday experience. You can call it hubris, but it is realistically grounded that we can, if we understand things deeply, do things that would look like magic. At every stage, as new things are discovered and understood, we have to see what the possibilities are. I do not think that there are upper limits to the amount of mischief that could be done. We have to be careful and serious.

The prospect of collider-induced doomsday scenarios made for some worthwhile debates over science, the nature of risk management, and public accountability, but they were major distractions from the work going on at accelerator labs. By 1999, the multi-million-dollar refurbishment of the Tevatron at Fermilab was well under way. The upgrade meant the machine was finally powerful enough to embark on a serious hunt for the Higgs boson. At CERN, the LEP collider had only one more year left to run before closing down. When it did, the Tevatron would be the only machine in the world with a chance of discovering the elusive particle.

9

The Gordian Knot

The year the world learned how it all might end was the year CERN decided to go for broke. In 1999, the LEP collider had run for a decade, surpassing all expectations. With only a year left before finally shutting the collider down, managers gave the green light to push the machine as hard as it would go. They had one goal in mind: to bag the Higgs boson before they pulled the plug.

The collider had already been upgraded to operate at twice the energy it started out with, but so far none of the collisions had turned up anything that looked much like the Higgs particle. All CERN could do was wring as much energy as they could from the aging accelerator and wait. If luck was on their side, the particle might show up.

Squeezing more energy out of a collider that is already running at full tilt takes creative thinking. There is no big dial you can turn up an extra notch, no call you can make to the power company for a smidgen more juice, and, with a machine like LEP, there is no users' manual to flip through for ideas.

The task of driving the collider beyond its design limit was overseen by Patrick Janot, a French experimentalist who had joined CERN in 1987. If Hollywood ever made a film about the hunt for the Higgs—and CERN has featured in questionable productions before—Janot would be a gift to screenwriters. In his mid-forties, he has the prerequisite good looks and brilliant mind. But more important is his attitude. Janot is uncompromising and doesn't hide his passion for particle physics.

Before you try and soup up the world's largest machine, you need to know what goes on under the bonnet. In 1999, the place to go and find out about LEP was Building 874. Across the road from the main CERN campus, the building was home to the LEP control room, the brains of the collider. From here, more than 3,000 kilometers of optic fibers reached out through the soil to every extremity of the machine. They ferried signals to and from the control room, where rolling shifts of technicians watched around the clock as a hundred thousand vital signs danced on their computer screens.

Janot spent a good deal of 1999 in the control room. He talked with the operators, listened to their discussions, and watched how they worked. He got to know the machine's physiology. "I was gathering all the information I could, to try and understand how to improve the way it was run," Janot recalls. "My mission was clearly spelt out to me by the research director. It was to bring LEP to the point where we would discover the Higgs, if it was there."

The Higgs is a tricky beast to snare. The particle is so unstable that it would survive for just a hundred trillionths of a trillionth of a second. That means that, even if the particle appeared at LEP, it would vanish again before you could blink. Particle physicists say "the Higgs doesn't fly," meaning it is so short-lived that you have no hope of seeing one directly in a detector. No sooner is it made than it decays, with a metaphorical puff, into other less exciting particles. A complicating factor is that at LEP the Higgs boson would never turn up alone. It would usually be made alongside a Z particle, which also decays into a spray of other particles. To find the Higgs boson, you would have to spot the tracks that all these particles made in a detector, and then work backward to see if any of them came from the elusive beast.

There's more to finding the Higgs particle than just waiting for the telltale signature of streaks and swirls to show up. With every Higgs hunt there is an added frustration. Other particles made in the machine can produce signals that look just like the Higgs boson. For example, as soon as LEP reached an energy of 182 GeV, which happened late in 1997, each collision was sufficiently energetic to produce two Z particles. Each of these could decay into two quarks, producing a total of four "jets" of particles. When a Higgs is made with a Z, they both decay into quarks, too, producing a similar pattern. The result is that the tracks left by the Higgs are hidden by debris from Zs and other particles.

In the summer of 1999, many physicists believed the Higgs mass lay somewhere between 100 and around 250 GeV. By driving LEP harder, they hoped to explore this territory, and plans were laid to boost the machine's performance into 2000, its final year of running. There were other good reasons for pushing to higher energies as well. CERN's next machine, the Large Hadron Collider, would struggle to find the Higgs if it was lighter than around 110 GeV.[1] In the LHC, the collisions would produce so much subatomic debris that it would be almost impossible to see the fingerprint left behind by such a light Higgs particle. CERN had to make sure it left no gaps where the Higgs might hide. There was also the competition to think of. The more ground LEP covered, the harder it would be for Fermilab's Tevatron collider to discover the Higgs particle first. Some scientists at CERN even had a motto: every extra GeV they could muster meant a year's extra work for the Tevatron Higgs hunters.

The first major tweak came in the form of a new refrigeration system.[2] Some of the units that accelerated the particle beams in LEP used superconductors, which were extremely efficient when kept cold enough. The new cooling system was better at bathing the accelerator units in liquid helium, chilling them to around −269 degrees Celsius. In outer space, you won't get colder than −270 degrees Celsius. With the system fitted, technicians squeezed around 5 percent more energy into the particle beams.

More modifications followed. Some old accelerator units that CERN had retired were dusted off, bolted on, and pressed back into service. Then, the beam trajectories were nudged around so that the huge magnets used to focus them gave them an extra kick of energy.

With every adjustment, Janot and the team pushed the machine closer and closer to its breaking point. To some CERN scientists, it was like being on the Starship Enterprise with an overworked, oil-smudged Scotty shouting from the engine room: "She cannae take any more!"

When everything had been done to tune up the ailing accelerator, the team decided to push the machine so hard it tripped out. In LEP, the accelerator units were driven by microwave generators called "klystrons." You'll find a klystron in your microwave oven, though it should be a little smaller. Usually, the collider had a safety margin that meant if one klystron tripped out, a couple were on hand to make sure the particle beams kept circulating. In the summer of 2000, the machine ran without any backups at all. If a single klystron tripped, the machine would break down.

It took Janot a while to convince the accelerator technicians to run the machine in such a precarious way. If the machine tripped out, it would be a long and painful procedure to get the beams up and running again. Even then, you could never be sure the beams you got would be as good as the ones you had just lost. It was tiring work for the accelerator technicians and frustrating for the detector scientists, who had minutes to grab what data they could before the beams inevitably went down again.

For two weeks that summer, in one of the most exasperating periods of the machine's life, LEP ran exclusively above 208 GeV. The klystrons were tripping every fifteen minutes, and it took at least half an hour to fill the machine up with racing particles again. The technicians and scientists were exhausted and frustrated.

At this point, Janot had a thought. He did a quick calculation and went into the control room. He told the operators that if they could get two consecutive runs to last more than forty-five minutes each, he would get naked in the control room.[3] Somehow, the next two runs lasted fifty-one minutes and an hour forty, respectively. "How did they do that? It was impossible!" Janot says. He had computed the odds and decided they were minuscule. To the relief of the control room staff, Janot backed down on this promise and paid them off with champagne instead.

The collider was straining. Every tweak seemed to further its transformation from a smooth-running precision instrument to an unpredictable wild beast of a machine. It would roar to life with impressive intensity only to collapse minutes later, as if from exhaustion. Down in the LEP tunnel, the radiation given off by the beams was starting to bake some of the machine's components, turning them an unhealthy shade of yellow-brown.[4]

On June 15, a Greek physicist, Nikos Konstantinidis, logged on to CERN's internal website for Aleph, one of the collider's four big detectors, to look at the latest results. Every morning, the detector automatically ran a computer program that sifted through the previous day's collisions and posted any events that looked interesting. That morning it had found something. Konstantinidis clicked on the event and started looking it over. It showed four clear jets of particles, each produced by a quark hurtling away from the collision. After a few more checks, Konstantinidis worked out that two of the quarks came from the decay of a particle with a mass of around 91 GeV—a Z particle for sure. The other two quarks came from something else, though, a particle around 114 GeV. It was too heavy to be a Z particle.[5] Konstantinidis went next door and talked to his colleagues. At their suggestion he ran some more checks. "As I was looking at it I was asking questions, querying it. It took me half an hour, an hour maybe. And the more I studied it, the more I checked it out, the more it looked like a Higgs. It was a beautiful, remarkable event," he said.

Impatient people don't go far in particle physics. The detector had picked up what looked like the aftermath of a decaying Higgs particle, but a fleeting glimpse means next to nothing in physics. You have to see the same thing over and over again to be sure it's not a trick of nature. In the quantum world, confusing things abound.

The rules for making discoveries in particle physics go as follows. Before you are allowed to shout "Eureka" and go running around the laboratory, you need to be absolutely sure your new particle isn't something boring that just happened to have popped up and muddled your data. How can you be absolutely sure? The particle has to appear so clearly and reliably that the chances of it being a statistical fluctuation in the data are less than one in a few million. You do this by comparing the events you see with the number you would expect if the particle didn't exist. Physicists use a statistical measure called standard deviation—denoted by the Greek letter sigma—to rank their confidence in potential discoveries. If you have a 3-sigma discovery, you can claim "evidence" for a new particle, but to qualify as a true discovery a signal of at least 5 sigma is required. It means the chance of the result being false is less than one in a few million. There is no real limit on the sigma scale; every increase means the result is more reliable.

At the end of June, the Aleph team met to discuss their latest results. Konstantinidis stepped up and talked through the event he had been working on. It stood out among all the other collisions. It was only one event, but gradually people started to wonder: had they finally seen a glimpse of the Higgs?

There is a tradition of secrecy among the teams of physicists that work on particle colliders. Each group studies its own collisions, does its own analyses, and does its best to keep its discoveries to itself. But in a lab the size of CERN, rumors spread fast. It wasn't long before the cautious excitement in the Aleph team became public knowledge. Patrick Janot summed up the feelings of many CERN scientists. "It's as though the years of restraint and frustration are finally being satisfied. But you're a scientist and you know that in statistics, things come and go. You say, okay, let's not lose ourselves. This is our chance. Let's get it."

The Aleph physicists were up against it. The one thing they needed was time, and time was in short supply. The LEP collider was scheduled to shut down in mid-September, leaving them only a couple of months to see if more Higgs-like signals might turn up.

In July, CERN held one of its regular meetings of the LEP Experiments Committee, where scientists and engineers on different parts of the machine gave progress reports and raised any problems. Spurred on by the Aleph signal, all four detector groups urged the management to run the machine for two weeks longer than planned.[6] CERN managers had built in a two-week reserve as a contingency and agreed to the run-on. The collider, they said, could operate until the end of September. A closed meeting for a handful of the researchers was set for the beginning of that month to review the situation.

The CERN teams worked around the clock. They tried new ways of analyzing the collisions in the hope of improving their chances of spotting a Higgs boson. The operators made sure the machine kept running at the very limit of its capability.

At the September meeting, CERN's director general, Luciano Maiani, and the director of research, Roger Cashmore, sat down with other senior managers and the CERN scientists to hear if there was any progress. Since June, the Aleph team had spotted two more collisions that produced patterns tantalizingly like the Higgs particle. They were just like the one Konstantinidis had seen on June 15. The strength of their signals ranked at 3.9 sigma—solid and encouraging, but not enough to claim a discovery. A second team on the Delphi detector had also spotted two collisions that might have created Higgs particles, but they were less certain. When the results from all four detectors were combined, the evidence for the Higgs particle was calculated to be 2.7 sigma.[7] By physicists' reckoning, it was not enough to qualify as evidence for the particle's existence.

On the basis of growing evidence, all four detector teams urged the CERN managers for a stay of execution. Officially, LEP had less than four weeks left to run. They asked for an extension of two months, meaning the machine would close for good in the first week of December 2000. In that time, the teams hoped to get even more Higgs signals. After the meeting, Tiziano Camporesi, head of the Delphi detector, said: "We don't want to be the scientists who go down in history as having missed the Higgs." Another CERN physicist, Chris Tully, was upbeat: "The Higgs is on the horizon."[8]

The CERN management made it clear they would only keep LEP running if there were "good prospects of transforming a tentative observation into a significant discovery."[9] After a long discussion, they decided on a compromise. They granted an extension until November 2.[10] To run on any longer, they argued, would delay engineering work on the Large Hadron Collider unnecessarily.[11] Even if the Higgs boson was there, and the machine kept running until December, the scientists were unlikely to make enough Higgs particles to reach the magic 5-sigma level of confidence needed to claim a discovery.

Rumors that a discovery was in the cards at CERN spread far beyond the laboratory. Six thousand miles away, a group of scientists had met for dinner on the island of Jeju off the southern tip of the Korean peninsula. They were visiting for a conference called Cosmo 2000, which covered particle physics and theories of the early universe. As the conversation turned to physics, Gordy Kane, director of the Michigan Center for Theoretical Physics, mentioned that CERN might have caught a glimpse of the Higgs particle. The discussion was interrupted by an unmistakable voice. It was Stephen Hawking, offering Kane a wager. The Higgs boson would never be found, at LEP or any other particle collider, Hawking said. The two agreed to a $100 bet on the matter.

Hawking's doubts about discovering the Higgs boson rested on work he had published in 1995, in which he predicted that "virtual black holes" could make it impossible to observe the particle.[12] Virtual black holes are curious theoretical entities. Scientists know that pairs of particles, such as electrons and their antimatter counterparts, positrons, can suddenly pop out of the vacuum because of energy fluctuations at the quantum scale. Virtual black holes are similar but are produced by fluctuations in spacetime. Their existence would be fleeting, but, Hawking argues, problematic enough to obscure the Higgs boson.

When Peter Higgs heard about Hawking's bet he wasn't impressed. Hawking, he said, merged particle physics and gravity in a way "no theoretical particle physicist would believe" was correct. "It was a bit of a cheek," Higgs said. "I am very doubtful about his calculations."

What began as a private disagreement erupted into a public row some time later.[13] At a dinner in Edinburgh, Higgs made an off-the-cuff remark about Hawking that was overheard by a reporter on the *Scotsman*. Higgs pointed out that it was hard to engage Hawking in debate and that the famous cosmologist wasn't challenged as openly as other physicists. "He has got away with pronouncements in a way other people would not. His celebrity status gives him instant credibility that others do not have," Higgs was quoted as saying. The comment brought a swift response from Hawking, who said: "I am surprised by the depth of feeling in Higgs's remarks. I would hope one could discuss scientific issues without personal remarks." The two later settled their differences, but not their difference of opinion about the chances of physicists ever seeing the particle.

In early October 2000, CERN held a long-planned celebration to mark the end of the LEP era. The machine had, so far, found no new particles, but it had measured the W and Z particles with extreme precision and put physicists' theory of matter, the Standard Model, on a surer footing. As the celebrations began, the CERN managers held another closed meeting with a small group of scientists to hear how the Higgs hunt was going. The situation hadn't changed much. The only possible glimpses of the Higgs particle were still from Aleph and Delphi, and the combined strength of data from all four detectors had been recalculated at 2.5 sigma. The agreed extension gave scientists the chance to boost this to a 3-sigma signal— enough, at least, to claim official "evidence" for the Higgs boson. The scientists were sent away and told to get their final results ready for November 3.

The CERN scientists had good reason to be cautious about the Higgs-like events they had recorded. All five showed what looked to be the Higgs boson and accompanying Z particle both decaying into quarks. While that was the most likely way the Higgs would show up, it wasn't the only way. Normally, you would expect to see that happen 70 percent of the time. Around 20 percent of the time, the Z would decay into neutrinos, which zip through the detectors without a trace. The rest of the time, the Z would break down into electrons or their heavy cousins, muons. It was suspicious that none of these other signatures, which should happen roughly one in three times, had been spotted.[14]

A few weeks later, Ross Berbeco, a physicist on another LEP detector called L3, was having a long Wednesday at the lab. In a last push before the final meeting, he was processing data the detector had collected over the previous two weeks. The clock on the wall said midnight. "I was hoping just to get it done, go home, relax, call my girlfriend," he said.[15] Once the processing was done, Berbeco took a look through the data. There was a Higgs-like event there, taken around 10 P.M. on October 16. It was a different kind of signal, the second most likely sort where the Z particle breaks down into neutrinos.

A number on the screen had Berbeco transfixed. It was calculated by the detector software and said how closely the collision debris resembled what should appear with a Higgs. The number could be anywhere between 0 and 1, with 1 being the best possible outcome. The number on the screen was 0.9995. Berbeco spent a few hours staring at the event and examining it. He went over his work to make sure he hadn't done anything wrong. At four in the morning he decided to call it a night. On the way out, he emailed his boss and suggested he take a look at the event for himself.

Word soon got around that L3 might have seen the Higgs particle. Three things made the event important. Calculations showed that if it was a Higgs boson, then it was around the same mass as the particle the Aleph detector had seen. It was independent confirmation from another detector, which allayed any fears that Aleph's signals were due to some quirk of the equipment. Third, the Higgs-like pattern L3 had picked up was different from those in Aleph. All of these things were exactly what the physicists would expect to see if the Higgs particle was really there.

The collider had only days left to run, but many scientists and engineers felt the Higgs particle almost brushing against their fingertips. The machine operators and detector groups quickly began working on an ambitious bid to persuade the management to keep the machine running for longer. And not just by a few weeks. They wanted to work on the machine over the winter break and push it to even higher energies over a six-month run in 2001. By refurbishing old accelerating units, they could drive LEP at more than 208 GeV, in spite of the tunnel shaking under the impact of civil-engineering work that was already under way to build LEP's successor, the Large Hadron Collider. If the plan went ahead and the Higgs boson's mass was around 115 GeV, as the experiments now suggested, the scientists had a good chance of turning what was a combined Higgs signal of 2.9 sigma into an indisputable 5.3-sigma discovery.

Scientists took the plan to CERN's managers, who went through it in detail with the LEP Experiments Committee. According to the statistics, the chances of the Higgs-like patterns being a fluke and having nothing to do with a real Higgs particle were only 0.2 percent. That said, the committee took a conservative stance. They decided the odds were no more than fifty-fifty that the Higgs boson was waiting for them at 115 GeV. What complicated things was that if the Higgs particle was only slightly heavier, at 116 GeV, then they were unlikely to make a formal discovery even if they did run the machine in 2001. According to the minutes of the meeting, there were "sizeable prospects for a Higgs discovery if LEP operates in 2001," but this optimism was qualified.[16] The minutes go on to add: "Even if the present events are due to a Higgs, there is roughly a 20 per cent probability that its mass is too high for a one-year extension to establish a discovery."

The scientists debated long and hard over what to do. The Higgs particle seemed so close, and the evidence seemed to be improving by the week. Switching off now would hand the initiative to the U.S. Tevatron. The Fermilab collider was scheduled to begin its highest energy run ever in spring 2001 and had a good chance of beating CERN to the Higgs before the Large Hadron Collider was built. The downside of running LEP in 2001 was that it would presumably delay the LHC, a vastly superior machine and the future of the laboratory. While many scientists lobbied CERN to press on with LEP for six more months, the Experiments Committee could not reach a formal consensus to put to managers.

That afternoon, hundreds of CERN scientists packed into the main auditorium to hear the first open talk on the latest in the hunt for the Higgs boson. Peter Igo-Kemenes, a physicist from the University of Heidelberg and a member of CERN's Higgs working group, took the audience through the evidence. It all pointed to a Higgs boson with a mass at 115 GeV. As he put a summary slide up, he paused and turned to face the room: "It's pretty exciting," he said. The audience erupted. They gave Igo-Kemenes a standing ovation, but the applause was for all the scientists and engineers who had achieved so much in the dying days of the accelerator. Igo-Kemenes ended his talk with a slide that requested the machine be allowed to run in 2001, adding: "The four experiments consider the search for the Standard Model Higgs boson to be of the highest importance, and CERN should not miss such a unique opportunity for a discovery."

The day before the talk, Patrick Janot had been working in his office when he took a call. It was from a senior scientist who was close to Director General Luciano Maiani. "Congratulations Patrick, you've won," the voice said. "One more year. Maiani has decided." It was unofficial, but Janot was jubilant. He had lobbied harder than anyone to keep LEP running. Janot went along to the talk in the CERN auditorium to see if Maiani was in the audience. He wanted to be there if there was an announcement. Instead, when Igo-Kemenes opened the floor for questions, Maiani quizzed him hard about the pattern the L3 detector had seen. He said nothing about running LEP for another year.

Soon after the talk, Janot got another call from the same scientist who had phoned him earlier. Maiani had done a U-turn. His plan to run LEP in 2001 had been scratched overnight. Janot was distraught. "I went from being delighted to being in grief. It was extremely difficult. It took me a year to recover. For a whole year I was missing something, something that I had worked so hard for, maybe discovering a particle of prime importance, and then being told: No. It's not for you." In an interview with the *New York Times*, Janot said: "When you have the possibility to draw a line in the history of mankind, you cannot miss that chance. And this is the choice in front of us."[17]

According to several CERN scientists, Maiani's change of heart happened after a series of one-on-one meetings with senior researchers at the laboratory. He talked to each of them in turn about the evidence they had for the Higgs particle. The night before Peter Igo-Kemenes gave his talk, when everyone else had gone home, a physicist on one of the detectors went back to see Maiani in private. The physicist told Maiani he was being fooled by the L3 event. It wasn't a proper Higgs signal at all. It was junk.

The strength of the L3 signal relied on the Z particle, which was made alongside the Higgs boson, decaying into two neutrinos. You don't see the neutrinos directly because they zip straight through the detector, but you can infer their presence from the "missing energy" they take with them, which is recorded by the detector. The process can be extremely difficult to distinguish from other processes, such as other particles decaying into high-energy photons or gamma rays.

Maiani worked out the expected background for the L3 Higgs signal, which gives a measure of all the other kinds of particles that could produce a false signal. The result, he recalls, made him "swear like a sailor," because it was so small. That meant a Higgs signal should stand out easily. "It made me wonder if we were just sitting on top of the Higgs, and this was at the very last moment. We discussed it and discussed it," he said. "But finally, I got convinced that on L3 they had no evidence of anything." Without L3's smoking gun, Maiani felt that pressing on was too much of a gamble.

A few days after Igo-Kemenes spoke to the packed CERN auditorium, the high-level research board at the laboratory met to make a final decision on the fate of the aging accelerator. The meeting took place on the top floor of the main CERN building. Many scientists at the lab saw it as their last hope. They gathered in another building a block away and looked over to where the meeting was going on. As the campus grew dark, the meeting stretched on. "We watched until the lights went out, because that was when we knew a decision had been made," one scientist said. The lights didn't go out until midnight.

At the meeting, the greatest worry was what impact running LEP for another year would have on the Large Hadron Collider. Lyn Evans, who was managing the LHC project, said the installation schedule would have to be revised. The machine might just be ready in time to run at the end of 2006, but that was a gamble. More likely, the first high-energy collisions would have to be put off until 2007. The cost of delaying and running LEP an extra year worked out to 120 million Swiss francs.

The financial penalty wasn't just the cost of plugging LEP in for another six months. By delaying the construction of the LHC, CERN would be breaking contracts with engineering firms that were already manufacturing components for the new collider. "It was clear those companies would sue us," Maiani said. "We would have had to pay their workers to do nothing for a year."

The research board next looked at whether the U.S. Tevatron collider at Fermilab would beat them to the Higgs particle if they closed LEP down. It was hard to be sure, but the CERN managers thought it unlikely that Tevatron would see the Higgs before 2007, by which time the LHC should be up and running. There was a chance, though, that the American collider would see some evidence before CERN's new machine was running at full energy.

What turned the meeting into a marathon was the discussion that went through each of the pros and cons of continuing with LEP. The Higgs particle would be a major trophy for CERN—and likely a Nobel Prize—and the evidence was growing exactly as if the particle was there. Turning LEP off now left the answer dangling and could gift the discovery to the Tevatron scientists. Others argued that the evidence for the Higgs boson wasn't strong enough to disrupt the LHC schedule and that the cost was too much to bear. Crucially, some argued that the Experiments Committee itself hadn't been able to reach a consensus on whether to extend the old machine. By the time the meeting drew to a close, the board was split.[18] The members couldn't agree on a recommendation to give the director general.

Patrick Janot had already begun a campaign to keep LEP in operation. He started a petition among CERN scientists in the hope of persuading management to reverse the decision. During the lobbying effort, someone at the lab sent him an unofficial construction schedule for the Large Hadron Collider. It suggested the machine was already facing delays, and so running LEP in 2001 would have almost no impact on its schedule. When Janot posted it on his website, he got a call from someone in the management team. "They told me to remove it immediately or I was fired," Janot said. Maiani admitted to commissioning the schedule, but claimed it only confirmed that trying to install the LHC in a shorter time frame was unrealistic.[19]

The day after the research board met, CERN scientists were dumbfounded to hear that a press release had been issued from the laboratory. It announced that LEP had been switched off for the last time.[20] Maiani hadn't received a straight recommendation from any of the senior scientific committees he'd consulted, so the existing plan to rip out LEP and make way for the Large Hadron Collider had to go ahead. He had cut the Gordian Knot. Staff scientists could hardly believe they had to learn such momentous news from a press release. The CERN staff association, which represented more than two-thirds of the workforce, didn't pull any punches: "Such a decision cannot be taken on the sly by a director general who no longer knows how to listen to the scientific community as a whole," it said.[21] The despondency ran so deep that it threatened to split the particle physics community.

The optimists at CERN saw one final chance for a reprieve. The press release said that CERN managers would not start dismantling LEP until they had the approval of the lab's member states, which were due to meet one Friday during the coming two weeks. News of that meeting leaked as soon as the member-state delegates filed out. In an email to his group, Dieter Schlatter, head of the Aleph team, said: "The sad news is LEP will be dismantled starting Monday. I have no official news yet, but from a member of the committee of council we learned that, as has been customary in recent committees, no consensus could be reached on any solution. Therefore the request by the director general to dismantle is valid." He signed off with these words: "Dreams do not always come true, but it is good to have dreams."

For Janot, this was the moment that CERN gave away its best chance of bagging the Higgs boson. The U.S. Tevatron scientists would pour all their efforts into finding the particle before the Large Hadron Collider was up and running, he said. "CERN will look ridiculous at having missed the opportunity, and the future of the CERN will be very dark," he told the British journal *Physics World*.

Back in Michigan, Gordy Kane wrote out a letter and put it in the post. It was addressed to Stephen Hawking at Cambridge University and enclosed a check for $100. In an exchange of emails a few years later, Professor Hawking said the bet that the Higgs particle would never be found didn't end with LEP. It carries on to Fermilab's Tevatron and CERN's Large Hadron Collider. "I think there's a good chance that virtual black holes will make it impossible to observe the Higgs, but, of course, if it is found, I will pay," he wrote.[22] On hearing this, Kane said he was confident he would one day get his money back.

Roger Cashmore, who was CERN's director of research when LEP shut down, took over as the principal of Brasenose College at Oxford University in 2003. He was at the heart of the discussions at CERN when the decision was made to close the LEP collider. "It was," he recalls, "a very, very fraught time."

The crux of the decision, he says, was that there was simply too little evidence for the Higgs particle too late in the game. Scientists argued that LEP should have run for another year because the laboratory's raison d'être was to discover new physics. If the Higgs boson had been found in 2001, scientists could have beavered away on the implications of its existence while heavy machinery installed LEP's replacement, the Large Hadron Collider. Ultimately, though, the decision could not be made on the value of the scientific discovery alone.

It later emerged that, under Maiani's leadership, an enormous hole had opened up in CERN's finances. The scale of the deficit, some 850 million Swiss francs (around $570 million at the time) wasn't made public until the autumn of 2001. Maiani apologized for taking so long to come clean about the problem, but said he was distracted by the excitement over the Higgs at CERN.[23]

"Had we run LEP for longer and not come up with anything and then discovered a bloody hole in the finances of the Large Hadron Collider, it could have been . . ." Cashmore stops and draws a finger across his throat. "It could have been the end of the subject. It was that dramatic. It was that heavy-duty. The Large Hadron Collider was not a given. People could have pulled the plug. They could have said we were irresponsible, spending money we didn't have, on an off chance. And it was an off chance. We could have ended up with nothing."

Cashmore goes on: "We had to make a decision: do we close down or do we keep going? I'm convinced it was the right decision to close down, but we made thousands and thousands of enemies. They would have turned around and stabbed us in the back. There were plenty of people who were absolutely clear that closing LEP down was the wrong thing to do." Back in Edinburgh, Peter Higgs fully supported the decision to close the collider.

In time, the hints of the Higgs particle at LEP faded. After the machine was switched off, scientists analyzed all of the evidence from each detector in more detail. When they were done, only Aleph's Higgs signals stood out against the background of other particles being created in the collider. The rest had vanished under more intensive scrutiny. The overall significance of the evidence had fallen to just 1.7 sigma.[24] The decision to shut LEP down would have been less traumatic if that had been known at the time. The final word from LEP was that the Higgs boson must weigh more than 114.4 GeV. The most likely mass from the suggestive Higgs signals was 115.6 GeV.

A year after LEP closed down, the defunct collider hit the headlines again in a story that made many scientists clutch their heads in disbelief. The British magazine *New Scientist* ran an article that suggested the Higgs boson didn't exist at all.[25] As if that wasn't bad enough, *The Times* picked up on the story. The headline for *The Times* read: "God Particle Disappears Down £6bn Drain." The piece wondered what the point of CERN's new project, the Large Hadron Collider, was if its most famous goal was a figment of scientists' imaginations. Physicists at CERN were apoplectic.

What happened sounds like a game of Chinese whispers. CERN was home to a group of scientists who formed what was known as the "Electroweak Working Group." As the name suggests, their interests lay in the physics behind the unification of electromagnetism and the weak force. The theory relied heavily on the Higgs mechanism. The group had plotted out the likelihood of finding Higgs particles of different masses, on the basis of other physical measurements, and the plot suggested the Higgs particle most likely weighed eighty times as much as a proton or 80 GeV.

The story broke when John Swain, a member of the CERN group, told a *New Scientist* reporter that LEP had ruled out more than half of the masses they thought the Higgs particle could have. Swain was quoted as saying: "It's more likely than not that there is no Higgs." That was one interpretation, but Swain claims he didn't mean to put it like that. As he explained later, hunting the Higgs particle was like looking for lost house keys. You search from room to room, and before too long you've checked more than half the house. You could, if you were so inclined, decide there and then that your keys didn't exist anymore. Most probably, though, you were just bad at guessing where you had left them.

A high-level group of CERN scientists wrote a letter to the editor of *New Scientist* to contest the story. They were astonished to read the article, they wrote, considering that "all our data are consistent and compatible with the existence of the Higgs boson, which remains one of the key issues for our understanding of particle physics."

Swain got a barrage of emails and phone calls. His colleagues were up in arms. The U.S. funding agency, the National Science Foundation, called to find out what was going on. "I had a lot of people get in touch who were very, very angry," Swain said. In an attempt at damage control, Swain drafted a letter to the editor of *The Times*. He tried to explain that the Higgs particle might merely be at the heavy end of expectations. He explained that even if it didn't exist, something like it did the same job and would be found by the Large Hadron Collider. The letter was never published. "I wrote to them, faxed them, called them, emailed, everything. I did the whole lot and got nowhere," he said. Eventually, Swain gave up and emailed the letter to all the physicists he could think of to explain what had gone wrong.

The story did nothing to boost scientists' trust in the media, but it gave a rare insight into the minds of some of the scientists hunting the Higgs boson. Swain got messages from researchers saying the Higgs had to exist, and that they had spent their professional lives searching for it. "They viewed it as an attack on a deeply cherished belief. It showed how much even scientists can come to believe in something," Swain said. "Belief can be helpful if you're searching for something, but you have to know that, no matter how much you like an idea, it might not be true. If we were absolutely certain we knew how things are, there would be no point in looking."

10

Chasing the Wind

Whoever named Chicago the Windy City must have been a glass-half-empty person. The architecture is handsome and historic, the museums and galleries world-class, and its vibrancy is unrivaled for nearly a thousand miles in any direction. That said, the name serves well as a warning for anyone visiting in the winter months, when a short walk to a café can become a survival situation if you are caught by the full brunt of the breeze. The wind is so chilling it makes the marrow in your bones flinch.

To get to Fermilab from downtown Chicago, you find Lake Michigan and drive in the other direction. After about an hour on the expressway, the city shrinks to nothing in the rear-view mirror and you pick up an access road that turns into the 7,000-acre campus where the laboratory is based. In the winter months, the entrance road weaves through a landscape of bare forests, frozen lakes, and yellowing switchgrass.

Further along, the route passes through the legs of a giant orange-brown sculpture that straddles the road like Talos waiting for the Argonauts. The 50-foot-tall structure was a favorite of the laboratory's first director, Robert Wilson. It has three legs that curve gracefully upward from the ground, but they meet over the road at different heights. The design means that when you approach from almost any angle the sculpture looks strained and off-kilter. Only when you lie beneath it and gaze up does its real beauty become apparent.[1] The symmetry is hidden. The three spokes spread out from the center to cut an imaginary circle into equal parts. Wilson named the sculpture *Broken Symmetry*, a concept that lies at the heart of the Higgs mechanism.

The hub of activity at Fermilab is Wilson Hall, a giant toast rack of a building from the side, but viewed from one end it could be a shrine to the Greek letter pi, or perhaps two giant hands coming together in prayer. The hall sits on one side of the Tevatron collider, which occupies a tunnel 4 miles around and 25 feet beneath the surface.

The Tevatron was built to smash protons into antiprotons inside two detectors, one called CDF (for Collider Detector at Fermilab) and the other named DZero (after its location on the accelerator ring). There are pros and cons to the design. On the upside, you can whip both beams up to speed with the same accelerator units because protons and antiprotons have the same mass and equal and opposite charges. An electric field that kicks a beam of protons one way will kick an antiproton beam the other. On the downside, antiprotons are messy to make and collect and are tricky to turn into well-behaved beams.

The machine saw its first collisions in 1985, but it wasn't until a decade later that scientists at Tevatron racked up their most prized discovery. In March 1995, the detector teams announced they had seen the top quark, the heaviest elementary particle known. The top quark weighed in at around 170 GeV, about as heavy as a gold atom. It was almost exactly where theorists predicted it would be.

Several cases of champagne later, the Tevatron was shut down for a planned refurbishment. The overhaul gave more oomph to the Tevatron's particle beams, raising the total collision energy to 1.96 TeV. The main improvement was to the detectors and accelerator equipment, which were upgraded to handle brighter beams.

The revamped Tevatron was on the verge of firing up again—with the Higgs particle in its sights—when hints of the elusive particle showed up at CERN's LEP collider. Robert Roser, one of the two scientists heading up the CDF team, has worked at Fermilab since the top quark days. When word got out that CERN might have found the Higgs, it was hard not to feel frustrated, he said. "We'd spent all this time upgrading to get a shot at the Higgs, and just as we were getting ready, someone's saying they might have it," he recalls. "Deep down, all we care about is the answer, but we have our pride and we are competitive. It's natural, when you've put blood, sweat, and tears into a job, that you want to be the ones who find it."

When LEP shut down empty-handed, the Tevatron was the only horse left in the race. The machine went into action in the spring of 2001, but it soon became clear something was wrong. Engineers struggled with the accelerator and for the first few years it underperformed. The lab took flak in the press, and relations became strained between the teams that ran the machine and the scientists on the detectors. The search for the Higgs boson was going nowhere, because the machine wasn't colliding enough particles.

The teething troubles at the Tevatron meant that at higher energies it was hard to keep the beams on a tight and well-defined orbit. That messed up the precision needed to line the beams up for head-on collisions. The entire ring of steering magnets had to be realigned with GPS devices and laser-tracking equipment. By early 2005, the accelerator was firing on all cylinders and steadily crashing enough particles to give scientists a hope of finally seeing the Higgs boson.

Like anyone who has a calling, particle physicists go where the jobs are. As high-energy facilities rise and fall on different continents, scientists migrate to wherever they have the greatest chance of finding something new in nature. With modern computer networks, some make the move a virtual one and analyze collision data from the comfort of their university offices. Others up sticks to follow the action.

John Conway is a case in point. An experimentalist at the University of California at Davis, he spent years at CERN with the Aleph team, the group that went on to see tantalizing hints of the

Higgs particle. Long before the excitement broke out, Conway returned to the United States to help revamp the CDF detector at Fermilab. A while later, in December 2006, he was back at CERN again, this time to arrange the delivery of some exquisite electronics designed to track particles inside the Compact Muon Solenoid, or CMS, detector for the still-in-progress Large Hadron Collider.

Conway remembers the trip to CERN that December for other reasons.[2] For the best part of the year, his team had played a waiting game. They were holding out for the CDF detector to gather a good chunk of collisions before sifting through them for signs of the Higgs particle. At the Tevatron, a collision between a proton and an antiproton could, in theory, create a Higgs that quickly decayed into two particles called "tau leptons," which are plump cousins of the electron.

Conway worked with physicists at Rutgers University in New Jersey on a software program that counted the number of collisions in the CDF detector that looked like a Higgs particle decaying into tau leptons. Other well-known particles produce practically identical signals, so the team knew their software would rack up plenty of false sightings. The only way to know if the Higgs boson was among them was to see far more than the other particles could possibly account for. Seeing nothing is no disaster: at the very least, it helps you work out where the Higgs particle isn't.

It was a Saturday morning when Conway heard that the CDF data was in and all the necessary checks and controls were done. He had the green light to "open the box," as physicists say, and check to see what, if anything, the detector had found. Conway sat alone in the atrium of Building 40 at CERN, where scientists on the LHC's Atlas and CMS detectors have their offices, and fired up his laptop. He tapped a few keys and pulled up the results. As the graph appeared, Conway felt the hair on his neck stand up. There were more tau lepton signals than expected. "It was like, Holy crap! What is that?" Conway recalls. If the Higgs particle was there, this is just how it would start to reveal itself.

Conway had to get back to Fermilab. He called the airline. They had nothing free the next day but found a seat on a Monday flight back to Chicago. He switched his ticket and fired off an email to the team at Rutgers to let them know what was going on. They were still asleep, but would get the message soon enough. Conway looked up at the spoked glass roof that topped the atrium. It was going to be a busy weekend.

The graph on Conway's laptop showed a clear bump that looked like the first glimpse of a Higgs particle weighing around 160 GeV. Conway knew it could be nothing more than a statistical fluctuation. The one thing particle physicists learn early in life is that a signal that looks exciting one day can vanish the next. What Conway needed to check were the odds of a bump that size appearing by chance if there wasn't a Higgs particle there. Physicists have a tried and tested way of calculating this, but it isn't quick. You run a computer simulation that banks the results from a staggering number of imaginary collisions that assume the Higgs particle doesn't exist. Then you look to see how often a bump like yours appears. If the answer is quite often, you can forget about shouting "Eureka!"

Conway pulled together the software he needed to run the simulation and added extra code to split the job among twenty different computers. With a push of a button, he beamed the commands to a bank of computers at the University of California that quietly hummed into action. Working in tandem, it would still take a few days to crunch the numbers and come back with an answer.

There was more to do. Conway had been invited to present the latest results of his Higgs hunt at a conference in Aspen the following January. That left just four weeks over the holiday season to get the new data checked out, nailed down, and approved for public release. The process was formal and thorough. Everything had to be written up and defended at two group meetings, where the intensity of questioning would be the closest many physicists get to a blood sport.

It was always a long shot that Conway's group had found the Higgs particle. More likely, it was a cruel but random blip in the data that would vanish as the CDF team analyzed more collisions. If that was the case, the bump in the graph was a pain. At the very least, Conway hoped to tell the Aspen conference that his team had ruled out certain masses for the Higgs, and so narrowed down where it must be. The blip made things a lot more uncertain. The bump in the data meant that a Higgs boson at 160 GeV couldn't be dismissed.

Conway got back to Fermilab on Monday and called an emergency meeting with three colleagues: Anton Anastassov, Cristóbal Cuenca Almenar, and Amit Lath. They had run checks on the data and were happy they hadn't made any stupid mistakes. The bump in the data was either just that, a random fluctuation, or it was a new particle. Right on cue, the job Conway had set to run on his university computers in California came back with an answer. The chance of the bump being a fluke of statistics was 2 percent.

The first group meeting, known as the "pre-blessing," went without a hitch, but word soon got around that Conway's group had an interesting result. More physicists showed up at the second group meeting and grilled the team with a barrage of tough questions. Eventually the group got their final blessing and the green light to discuss the results in Aspen a few days later.

At the conference, Conway talked through the Higgs search at CDF. He explained that a bump in the data made it impossible to rule out a Higgs particle at 160 GeV. He put up the same graph that had sent a chill down his neck at CERN that Saturday before Christmas, so the physicists in the audience could see for themselves. "It was a null result and I said so. I told them we have a small bump in the data and it is either a statistical fluctuation or it's the beginning of a signal, and you know how much stock to put in that," Conway says.

Despite Conway's caution, rumors that the team might have caught a glimpse of the Higgs particle spread like wildfire. "Within twenty-four hours I was getting emails and phone calls about it," Conway recalls. "They all wanted to know, had we seen the Higgs?"

The community had good reason to be excited. If the bump in Conway's data was caused by Higgs particles appearing and decaying into tau leptons, it had profound implications for the underlying laws of nature. The kind of Higgs particle Conway was hunting wasn't your common or garden variety. It was one of a quintuplet predicted by supersymmetry theory. In a supersymmetric universe, every particle we know has a heavier opposite number that lurks unseen in the shadows. The theory is mathematically beautiful but deeply counterintuitive. Proving it—and the discovery of a supersymmetric Higgs ranked as a good start—would pull physics out of the doldrums by finally opening the door to a world beyond the Standard Model.

For theorists who see beauty as truth, Nature should be ashamed if she didn't embrace supersymmetry. As is so often the case in physics, the importance of the theory hangs on its ability to unify. Maxwell unified electricity and magnetism, and in doing so explained light. Einstein unified space and time and showed that energy and mass were interchangeable currencies. Supersymmetry is arguably more ambitious still and unifies matter with the forces of nature.

The supersymmetric universe is a Scrabble player's dream.[3] Every force-carrying particle is paired off with a hypothetical matter particle. The photon, which carries the electromagnetic force, teams up with the "photino." Likewise, the gluons that carry the strong force inside atomic nuclei pair up with particles called "gluinos." Thanks to the force carriers of the weak force, the W and Z bosons, we get "winos" and "zinos." The arrangement works the other way too. Every matter particle we know of is partnered with a hypothetical force-carrying particle. The electron pairs with the "selectron" and the quark with the "squark." The Higgs boson gets its own superpartner too, the "Higgsino."

At first glance, supersymmetry seems like a mischievous ruse to make the world more complicated than it is already. But there are compelling reasons to suspect the theory might be an accurate description of the underlying structure of nature. Aesthetic and linguistic point-scoring aside, supersymmetry resolves some embarrassing cracks in the Standard Model that suggest it must be superseded by something better.

Take the Higgs particle. In the quantum world, the Higgs is easily influenced by virtual particles that pop in and out of the vacuum. These fleeting particles contribute to the mass of the Higgs boson itself and could plausibly make it spectacularly overweight. If you tot up the extra mass these virtual particles can bestow on the Higgs, its weight swells to more than a million billion times the value that experiments suggest. Physicists call it the hierarchy problem. A heavy Higgs particle cannot do the job it is designed to do, that is, breaking the electroweak symmetry. If the Higgs is a leaden couch potato, scientists will be forced back to the drawing board to find the origin of mass.

It is possible that the laws of nature are so finely tuned that the weight-gaining effects of virtual particles largely cancel out and the Higgs particle stays light. But the prospect makes scientists uneasy. For that to happen, the properties of the particles and forces in the Standard Model would have to be tuned to a precision of at least fifteen decimal places. Why would the laws of nature be so exquisitely sensitive to the most minuscule changes? Supersymmetry offers a way out. In the supersymmetric quantum world, superpartners cancel out the mass that the Higgs boson gains from virtual particles. If a W boson pops into existence and makes a passing Higgs particle heavier, the W's hidden twin, the wino, is on hand to take the weight off again.

Supersymmetry ties up another loose end left dangling by the Standard Model. Physicists believe that, in the immediate aftermath of the Big Bang, the forces of nature we see around us today were combined into one superforce. When the temperature of the universe cooled to around a million billion degrees, the Higgs field switched on and separated the electromagnetic and weak forces. But the strong force that holds quarks together inside protons and neutrons split off even earlier, when the cosmic thermometer read ten billion billion billion degrees. Herein lies the glitch. If you wind the clock back to the beginning of the universe, scientists think, the different forces should eventually meet up and merge into one, but with the Standard Model that doesn't happen. They come close, but they never become equal. In supersymmetric theories, this messiness is tidied up and the forces all meet at a single point. What thrills physicists even more is that supersymmetry shows a way of unifying these forces with gravity, which is completely ignored by the Standard Model.

There is also a chance that supersymmetry will give scientists a handle on one of the most baffling puzzles they face. It is something of an embarrassment that in the twenty-first century, we cannot explain what 96 percent of the observable universe is made of. Observations of our cosmic neighborhood have led scientists to notch up 70 percent of whatever's out there to dark energy, the mysterious force that is driving the expansion of the universe. No one knows what dark energy is, and finding out is a major goal in physics. The Standard Model describes but 4 percent of the matter we can see in space. The rest, about a quarter, is dark matter. The name is intentionally unilluminating. Dark matter doesn't shine or radiate heat, and you can't see it by bouncing light off it. Many theorists think dark matter is made of supersymmetric particles called "neutralinos," the lightest kind the supersymmetry theory predicts.

Not everyone is a fan of supersymmetry. Critics challenge aficionados with a simple question: If the universe is supersymmetric, where are all the other particles? Why haven't we discovered selectrons by now, or photinos or winos? The usual retort is that in the real world supersymmetry is broken, making the masses of the superpartner particles heavy. This defense is not so outlandish. The Higgs field breaks the symmetry underlying the electroweak force, making the W and Z bosons heavy and leaving the photon massless. Supersymmetry could be broken in a similar way, making superparticles heavier than normal matter.

At Fermilab, Conway was searching for a breed of Higgs particle predicted by what is called the "Minimal Supersymmetric Standard

Model" (MSSM). According to the theory, there are five Higgs particles in all, and they all have different weights and supersymmetric partners. Three of the particles are neutral, and two of them are charged.

When Conway wrapped up his talk at Aspen, he headed for the gleaming white mountains that make the town one of the most stunning ski resorts in the world. On the slopes, he met up with Greg Landsberg, a physicist at Brown University in Rhode Island. Landsberg worked on Tevatron's DZero detector, CDF's opposite number at Fermilab. He told Conway the DZero team was about to announce some new results that he might find interesting. The conversation stuck in Conway's mind for the rest of the afternoon. Had DZero seen a hint of the Higgs too? That evening, Landsberg came clean. The DZero detector hadn't seen anything that looked like the Higgs. Where Conway's CDF team had a rising bump in their data, the DZero team had a dip. The chances of the Higgs being there were looking slimmer than ever.

There was little to do but wait. Conway's team needed to analyze more collisions. If the Higgs particle was there, the bump would grow over time. If it wasn't, then the bump would settle down and eventually disappear. The team decided to record more collisions over the next six months or so and take another look at the end of the summer. Until then, they were in limbo.

Few physicists realized at the time that the potential sighting of the Higgs particle was about to challenge some of the most treasured but unwritten laws of science—not the physical laws that govern particles and forces but the social laws that act on scientists and the way they behave. Traditionally, hints of impending discoveries in particle physics circulated on the rumor mill, through word of mouth and email, and only rarely broke out into the wider world. The case of Conway's Higgs-like bump changed that forever.

At the turn of the millennium, tools became widely available for anyone to set up a blog on the Internet. Over the next few years, scientists started blogging. Unsurprisingly, they focused on issues that scientists cared about, but some talked about new findings in the pipeline that weren't yet official. That was a watershed moment. Science traditionally worked to rigid guidelines where new results were made public only when they had been peer-reviewed. The process was designed to weed out glaring mistakes and ensure that only sound studies make it into the scientific literature.

A few weeks after speaking in Aspen, Conway made his writing debut for the U.S. blog Cosmic Variance. In two colorful pieces, he described the excitement at finding the Higgs-like bump in his data, but made clear that it was almost certainly a statistical fluctuation. The response to the pieces was enthusiastic. Scientists posted comments to wish Conway luck or asked for technical clarifications. Others were just interested. One reader asked whether the Higgs particle might be of any use when scientists finally discovered it, prompting Conway to recount how J. J. Thomson, who discovered the electron in 1897, suspected it would never find a practical application. Conway went on: "In the case of the Higgs, there is no immediate practical use. But it could lead us to new theoretical and experimental breakthroughs that could transform technology in ways we cannot imagine, any more than Thomson imagined our own modern electronic age."

Conway wasn't the first to blog about a Higgs-like signal at the Tevatron. A week earlier, Tommaso Dorigo, another CDF physicist, wrote a blog that also mentioned the bump.[4] Dorigo emphasized that it was probably a meaningless fluke of statistics, but he went on to discuss the possibility of it being a new discovery in the making. He even cited another analysis of collisions at CDF that seemed to support the existence of a Higgs particle weighing 160 GeV.

Problems started when the mainstream media got involved. That March, *New Scientist* picked up on the blog posts and ran an article on the tantalizing signals that had particle physicists so excited.[5] A piece in *The Economist* followed soon after. A few months later, the *New York Times* covered the story. The articles had a startling effect on the particle physics community, considering that none of the pieces had claimed that the Higgs boson had been found, and all had made clear that the most likely explanation was a statistical blip. It was as if, simply by covering the story, the media had overstated physicists' claims.

The backlash was immediate. Scientists criticized the media for exaggerating the story, even when the articles under fire had been run past physicists for accuracy. It's not uncommon for that to happen. A tentative finding couched in complex reservations takes on a bolder stature when passed through a news reporter. That transformation alone is hard for some to stomach, and scientists who are quoted can easily be made to feel complicit. The media weren't the only ones being blamed, though. Physicists criticized Conway and Dorigo for writing about such ambiguous results on their blogs. The underlying sentiment was that the bloggers had crossed a line that made many in the community uncomfortable. "A lot of our collaborators got really upset with the idea that we'd write about science results in such an informal medium as a blog," says Conway. "I thought it was just a bunch of nerds reading our stuff. When the media picked up on it, that was an education." Dorigo went so far as to publish an open letter of apology on his blog. "I apologise because I know some of you feel betrayed," he wrote to the CDF team.

Some Fermilab scientists worried that blogging was an intrusion. They wanted to go along to meetings and speak their minds without fear of what might turn up on a blog that night. Some objected that results were being made public before they were formally peer-reviewed, but, in this case, all of the results had been through two levels of internal review and approved for public release. External peer review was unlikely to find any glaring flaws in the work. Perhaps the most sensitive issue was that in a blog written by one person it is hard to fully acknowledge that the results are the work of sometimes hundreds of people. Blogs seemed to go against the spirit of teamwork by inevitably emphasizing the individual.

The blogging row broke while Robert Roser was co-spokesman for CDF, alongside another physicist, Jacobo Konigsberg. Roser never thought Conway or Dorigo had overstepped the mark with their blogs, but he was frustrated with some of the media coverage. "This was the first time that articles were being written not from a research paper but from a blog," Roser says. "The reaction among physicists here was right across the map, from any press is good press, to how could we have let this happen?"

The two spokesmen got together and drew up rules of engagement to make clear what was and wasn't acceptable behavior for bloggers. They urged them not to air dirty laundry in public, to respect their colleagues, and not to blog about results before they had been formally blessed. In principle, whoever is responsible for doing the lion's share of the analysis gets to make a result public first, but technically, as soon as a result is approved internally, it can appear anywhere. "The fact is," says Roser, "high-energy physics is still getting to grips with how to deal with blogs."

Late that summer, the members of Conway's team were ready to analyze their latest hoard of collisions at the Tevatron. If the bump remained or had grown at all, it would be a sure sign that the Higgs was there and that we live in a supersymmetric universe. When all the checks were done, Conway ran the analysis. The graph appeared a second or so later. Nothing. The bump had vanished. Even though Conway had expected as much, it was a disappointment. Conway's blog summed it up: "So the quest for this beast continues. Mother Nature is a big fat tease!"

The year had been a roller-coaster ride. Thinking it over some time later, Conway recalled the feeling he had that Saturday morning at CERN when, all alone, he had seen the bump for the first time. It is a feeling that drives many people to do science. "You have this hope that someday you'll see something that is genuinely new, that no one else in the world has ever seen," he said. "You want to make a discovery."

At Fermilab, there are two detectors that physicists use to hunt for the Higgs particle. John Conway's team searched for evidence of the elusive boson amid collisions recorded by the CDF detector.

Other groups use the DZero detector. One of the spokesmen for the DZero collaboration is Dmitri Denisov, a Russian-born scientist who was educated in Moscow by some of the country's most respected physicists. Denisov was at Fermilab when the top quark was discovered in 1995. He is not a glass-half-empty man. He believes that if the Tevatron runs for a few more years, the machine has a 50 percent chance of finding the Higgs particle—provided it is not too heavy. "We are enthusiastic," he says. "We know how to make discoveries."

On September 19, 2008, Lyn Evans was deep in conversation at CERN's personnel office when his cellphone rang. It was the control room. He'd better come fast. Something had gone badly wrong. Evans raced across the campus and into the building, where technical staff were in the final stages of getting the Large Hadron Collider ready to crash particles together. He couldn't believe his eyes. Alarms were flashing everywhere. The vacuum system was shot, countless magnets were knocked out, and sensors showed a vast cloud of helium gas spreading quickly around the tunnel that housed the giant accelerator. The incident had triggered an emergency shut-down.

Evans called a crisis meeting. They had to get someone down into the tunnel to see what had happened. The helium cloud was bad news. As the gas spread, it forced air from the tunnel, making it impossible to breathe. They called the fire brigade, who donned gas masks and air tanks and went down to investigate. A few hours later, the helium had cleared enough for CERN's own engineers to go in and inspect the damage. The scene was one of carnage. Giant magnets were ripped from their concrete anchors, connecting equipment was crumpled or torn, and ventilation doors had burst open. The wreckage was covered in soot and ice and flecks of molten metal. "We were in trouble," Evans says.

Nine days earlier, on September 10, 2008, Lyn Evans couldn't have been happier. The world's media had descended on CERN for what they merrily dubbed "Big Bang Day," the inaugural switch-on of the Large Hadron Collider. For Evans, this was the culmination of fifteen years' work to design and build the world's most complex machine. He wasn't the only one glad to see that day arrive. For the 6,000 or so scientists at CERN, that day marked the end of eight long years with no particle accelerator and no collisions to study.

Shortly after 9 A.M., enormous screens set up around CERN's Globe building cut to Evans in the LHC control room. He was sitting in front of a bank of computer screens in jeans and a striped short-sleeved shirt. Jokingly, he started a countdown to lift-off, or, in this case, to the first lap of a proton beam in the collider. "Five, four, three, two, one!" Nothing happened. Then, a bright spot flashed on one of the screens. "Yes! Yes!" cried Evans. The spot was caused by particles hurtling down the beam pipe. The machine had cleared its first major hurdle. A few hours later, protons had been sent both ways around the machine's 27-kilometer-long racetrack. The Large Hadron Collider was working better than anyone dared hope. It was open for business.

As with earlier machines, there were people who thought the Large Hadron Collider was too dangerous to turn on. Walter Wagner, the retired radiation officer who had failed to get an accelerator in New York closed down ten years earlier, filed a lawsuit in Hawaii citing fears that the machine might destroy the planet. Or, perhaps, the universe. Wagner sought a restraining order until the collider was proved 100 percent safe. The so-called doomsday suit was thrown out of court after lawyers for the federal government called its claims "overly speculative."

A few weeks before the LHC was switched on, Sean Carroll, a theoretical physicist at the California Institute of Technology, drew up a shopping list of new physical phenomena the collider might find and published it on the Cosmic Variance blog.[6] At the top of his list was the Higgs particle. Carroll thought finding the Higgs boson was a 95 percent certainty. "The Higgs is the only particle in the Standard Model of particle physics which hasn't yet been discovered, so it's certainly a prime target for the LHC, if the Tevatron doesn't sneak in and find it first," he wrote, adding: "There is almost a guarantee that the Higgs exists." Next on Carroll's list, with a 60 percent probability of being discovered, was supersymmetry. The chances of finding nothing new at all, perhaps the most terrifying prospect for physicists, came in at 3 percent. Incidentally, Carroll put the odds of the collider creating a stable black hole that eats the Earth at 10^{-25} percent. That's one-tenth of a million billion billionth of 1 percent. "So you're saying there's a chance?" Carroll quipped, with a nod to the 1994 movie *Dumb and Dumber*.

The detectors that scientists use to sift new physics from the debris of particle collisions inside the LHC are technological masterpieces. By far the largest is the aptly named Atlas detector, from "A Toroidal LHC ApparatuS," which is so big it would barely fit in an Olympic-sized swimming pool. The 7,000-metric-ton Atlas and the not-so-small CMS detector were designed with the Higgs particle in mind. They might see far more, though, such as exotic particles of dark matter or extra dimensions.

The Higgs boson might pop into existence in any of several ways inside the Large Hadron Collider, but scientists predict the most likely route to be when two gluons—the particles that bind quarks together inside protons—slam together and fuse. The energy released in the collision would theoretically create a Higgs particle with ease, though it would decay immediately afterward. If the Higgs is light, weighing in at around the same mass at which CERN's LEP collider saw intriguing Higgs-like signals, it will be easiest to see as it vanishes in a burst of gamma rays. If it's a bit heavier, say, more than 130 GeV, scientists will look for streaks left by four leptons, the family of particles that includes the electron. Finding the Higgs particle relies on plucking these signals from the subatomic detritus that swamps collisions in the LHC.

A little way around the collider from the Atlas experiment is the LHCb detector. Here, scientists are trying to settle a question that Paul Dirac raised in his Nobel lecture in 1933. When the universe was created in the Big Bang around 14 billion years ago, there were supposedly equal amounts of matter and antimatter. But what happened to all the antimatter? Are there antimatter stars shining in antimatter galaxies? The LHCb detector is designed to capture particles made up of bottom quarks, a heavy type of quark that should help unravel why normal matter appears to have won out.

For a month every year, CERN technicians will clear the LHC of protons and fill it with lead ions for experiments that center on the Alice (A Large Ion Collider Experiment) detector. When these ions are made to collide at close to the speed of light, they generate temperatures more than 100,000 times those found at the center of the sun. Under these conditions, the ions will melt into quark-gluon plasma, a bizarre form of matter thought to have existed in the primordial universe. By watching how it behaves, scientists on the Alice experiment hope to unravel how it gave rise to the particles we see in the universe today.

According to mathematicians in Russia, we might hear that the Large Hadron Collider has created Higgs particles by an unlikely route. Two groups in Moscow, led by Irina Aref'eva and Igor Volovich at the Steklov Mathematical Institute and Andrey Mironov at the Lebedev Physics Institute, think the Large Hadron Collider might create time machines.[7] Not the kind that H. G. Wells dreamed up in 1895, which looked like Santa had modified his sleigh with an antique chesterfield and a parasol. These time machines would be tiny wormholes, or distortions of spacetime, that could, in principle, allow particles, if not people, to travel back from the future.

The idea sounds outlandish. And it is. But the mathematicians argue that under certain conditions, two particles could slam together head-on in the LHC with such energy that they would cause shockwaves that might just distort microscopic regions of spacetime. Most physicists think it would take unfeasibly large amounts of energy to make a wormhole, but others put such a feat within the LHC's reach.

If Aref'eva and her fellow mathematicians are right, collisions in the LHC could create what physicists call a "closed timelike curve" in which time loops back on itself. That would mark a kind of year zero for time travel. Nothing from the future could travel back in time until the wormhole was created, and, even then, it could go back only as far as the moment the time loop appeared. According to the theory, which is highly speculative, some future generation might send a message back to CERN via a fleeting wormhole. Only slightly more likely is that scientists working at the laboratory will see subatomic debris from particle collisions disappear as the debris falls into a time loop. The Russians make a virtue of this, because it would be experimental proof that wormholes and time travel were possible.

The incident at 11:18 A.M. on September 19, 2008, put all hope of new discoveries at the Large Hadron Collider on hold.[8] The failure that had sent alarms flashing in the CERN control room was no minor glitch. First, the machine had to be made safe to work on: there were 15-metric-ton magnets wrenched half off their supports in the tunnel, a precarious danger. Then the real job began—to clean up the mess, piece together what had gone wrong, and make the necessary repairs.

The explosion happened as CERN engineers were firing up the last big power unit that supplies the collider. Inside the machine, enormous electrical currents circulate in superconducting magnets to create fields that steer the particles around the collider circuit. The magnets are connected to one another by superconducting wires and, to operate properly, must be bathed in superfluid helium at −271 degrees Celsius. That day, one connection out of 10,000 warmed up and lost its superconducting properties. The fault caused an enormous spark to arc through the helium vessel surrounding one of the magnets. The spark punctured a hole in the vessel lining, allowing helium to pour into the next part of the magnet housing. Emergency safety valves flipped open to allow the helium to escape into the tunnel, but they were quickly overwhelmed. The pressure built up inside the magnets and wrenched them from their mountings, in some cases tearing the concrete floor of the collider tunnel.

The damage was staggering. The sparks created by the short circuit had enough energy to melt nearly half a ton of copper. Six metric tons of helium were released, producing a pressure rise that damaged fifty-three huge magnets. More than half a kilometer of the machine was covered in a thin layer of soot. "To see this at the very end, when we were commissioning the last big power supply, was a real kick in the teeth for everyone," says Evans.

Engineers quickly identified the cause of the fault. The superconducting wire, which should have been brazed under a weight of 2 metric tons, hadn't been brazed at all. No wonder it had warmed up and vaporized. As engineers got to work replacing the damaged magnets with spares, technicians were sent in to check the other 10,000 or so connections. What they found wasn't heartening. Around half were dodgy and would have to be repaired before the collider was safe to switch on again. Another job was to fit an early-warning system and more safety valves around the machine so that, if a similar accident did happen in future, the helium couldn't build up and cause so much damage. The repairs closed the collider for more than a year and cost the laboratory £24 million (40 million Swiss francs).

It wasn't the end of CERN's woes. While working on the collider, engineers found further problems with some of the magnets that weren't related to the accident. Before the magnets can run safely at full strength they have to be "trained" by running them at increasingly high currents. With each stage of training, the magnet can generate stronger and stronger fields. To their dismay, engineers discovered that forty-nine magnets had somehow lost their training after being installed—but they had only checked part of the machine. The final total was likely to be much higher. With no prospect of retraining the magnets any time soon, CERN scientists had to accept that when the Large Hadron Collider was ready to switch on a second time, it could not run safely at full power.

The setbacks at CERN were a serious blow for particle physicists. In its history, the Large Hadron Collider endured delays lasting years along with cost overruns and catastrophic accidents. Talking through the events a month before the repairs were finished, Evans was circumspect. "We were all really down, but you cannot dwell on it," he said. "You have to remember. Nobody had ever built anything like the Large Hadron Collider before."

The year 2009 marked the forty-fifth anniversary of the Higgs particle's birth, at least in the theoretical equations written down in Peter Higgs's notebook. Physicists had been on the lookout for the particle ever since, and they had hunted for it in earnest since the 1980s. The fact that it had not shown up yet was starting to look a little suspicious. They had looked for it in the wrong places. It seemed to appear and then slip from their grasp. In the minds of most physicists, that was sheer bad luck.

Holger Nielsen, though, thought it was fate. Nielsen, a theoretical physicist at the Niels Bohr Institute in Copenhagen, is regarded as one of the founders of string theory, which casts all the particles in the universe as microscopic strands of vibrating energy. In league with another theorist, Masao Ninomiya at Kyoto University in Japan, Nielsen proposed a mischievous theory that offered an explanation as to why the Higgs had never been found.[9] Nature herself, or even "God," was conspiring against it. Nielsen's theory claimed the Higgs particle was so despicable to Nature that any machine capable of churning the particles out was destined to be sabotaged from the future. Nielsen likened it to what is known in time travel circles as the "grandfather paradox," where a person from the future travels back in time and kills his own grandfather. The theory connects the present with the future in such a way that the future events can ripple back in time and influence what scientists do today.

In a helpful gesture to CERN, Nielsen and Ninomiya suggested the management hold a card game to find out whether a malign force from the future was to blame for their misfortune in attempting to find the Higgs boson. The game was simple. You take a million playing cards and write on one of them the message: "Close LHC." If one of the managers picked that one card, it could be a Higgs trying to tell them something from the future. Or so Nielsen suggests.

The theory met with a cool response from other physicists. Some wondered if it was a spoof. The Italian theorist Tommaso Dorigo, who blogged about hints of the Higgs at the Tevatron, ran a digest of the paper on his website under the title "Respectable Physicists Gone Crackpotty." In the post, he went on to lay his own cards on the table, writing: "It's very sad to see some valuable minds writing such a pile of unmitigated bullshit."

Nielsen admits he doesn't really believe the theory, but in his articles, he talks playfully of the evidence to support it amid the history of the Higgs hunt. The Superconducting Supercollider was scrapped by the U.S. Congress in 1993 when it was only partially built. The Large Hadron Collider shut for a year after the catastrophic helium leak. Both of these setbacks, he says, make sense if "God" finds the Higgs particle abhorrent. "This 'God' is trying to avoid them in general," Nielsen said in one interview. "He rather hates Higgs particles."

11

Hidden World

Peter Higgs is taking a paddle in the sea and giving a double thumbs-up to the camera, wearing nothing but a light-green "mankini," an eye-watering garment that resembles a thong with shoulder straps. Sunlight glints off the water behind him and the chrome and glass fittings of the millionaires' yachts sparkle as they bob about in the background.

The giant "Photoshopped" image, which is propped against the wall of John Ellis's office at CERN, all too realistically melds the head of Peter Higgs with the body of Borat, a character from Kazakhstan created by the television comedian Sacha Baron Cohen in 2006. "Borat was the theme of our Christmas revue some time ago," Ellis says. "A student of mine made it. He's incredibly good at that sort of thing."

Ellis is one of the most prolific and wide-ranging theoretical physicists in the world, having authored nearly nine hundred scientific articles. He told physicists how to find the gluon, the particle that transmits the strong force; he suggested that the lightest particle predicted by supersymmetry was part of the curious "dark matter," the invisible cosmic substance that makes up 25 percent of the mass of the universe; and he showed scientists how the Higgs particle might appear in their experiments. Ellis, who has more than a passing resemblance to J. R. R. Tolkien's congenial wizard, Gandalf, was a student at Cambridge University for the last of Paul Dirac's lectures on quantum theory in the 1960s and came to CERN in 1973. Ellis worked in an office down the hall from where he is now when he started the hunt for the Higgs particle in the most tentative way in 1976. That year, he and fellow theorists Mary Gaillard and Dimitri Nanopoulos published the first detailed profile of the Higgs boson and how it was likely to reveal itself in particle-collider experiments. "Very few people cared about the Higgs boson in those days," Ellis recalls. "I never thought about how long it might take to find."

In the years since Ellis's paper appeared, scientists seem to have come achingly close to discovering the elusive particle. Though physicists have yet to lay their hands on the prize, the hunt at CERN and at Fermilab's Tevatron accelerator near Chicago has at least narrowed down where the particle might be hiding. CERN's last atom smasher, the LEP collider, said the Higgs boson must weigh more than 114.4 GeV. In July 2010, the Tevatron collider had ruled out a Higgs particle between 158 and 175 GeV. By combining other experimental results, the most likely mass of the Higgs particle comes out at between 115 and around 130 GeV.

In February 2010, the LHC was switched on for its first serious attempt to find what scientists at CERN simply call "new physics." The CERN management decided it was too risky to run the machine at full power, so for about two years straight it will operate at only half its capability, meaning that each of the accelerator's beams will be whipped up to an energy of 3.5 TeV before they are steered into collisions inside the detectors. Near the end of 2012, the collider will close for twelve to eighteen months to give engineers time to do repairs and install a safety system. The safety measures will protect the machine from damage should things go awry at higher energies, and only once this work is done will the LHC run at full power.

The decision to operate the LHC at half its design energy was not made easily, but CERN could not afford another crippling accident like the one that shut the machine down for a year in 2008. No collisions mean no data, which is frustrating for physics and devastating for up-and-coming students trying to earn Ph.D.s. But the decision has consequences. Many physicists concede that the machine may not make a concrete Higgs discovery before 2015.

It is entirely possible that the Higgs boson has already been created in particle collisions at the Tevatron and the Large Hadron Collider, but in such small numbers as to go unnoticed. A research paper published by Fermilab scientists in February 2010 illustrates the point.[1] A graph in the article shows how closely—or not—the results from the American collider fit with there being a Higgs particle of various masses. Part of the graph is consistent with, though not proof of, a Higgs boson that weighs 115 GeV. That is right where CERN's previous collider saw what seemed to be a glimpse of the Higgs boson in 2000. "Us theorists are having a bit of an argy-bargy about it," says Ellis. "That could be it, right there. They might be seeing the Higgs particle already."

There is a big difference between a hint and a discovery. As of the beginning of 2010, the U.S. government is expected to run the Tevatron collider until the end of 2011, around the time the LHC will close for repairs.[2] Most physicists think that is too little time to discover the Higgs particle for sure. The best the Tevatron scientists can hope for, they say, is the kind of excruciating glimpse of the Higgs particle that threw CERN into a frenzy months before the LEP collider closed down for good in 2000. "Fermilab is giving it a bloody good shot, a better shot than anyone imagined, but I just don't think they have the mojo. They won't have enough collisions," says Ellis. "If they are very lucky, they may be able to claim some evidence, but not a discovery."

Building 40 at CERN is the de facto headquarters for the LHC's Atlas detector team. With more than 3,000 physicists belonging to the group (most work from their home universities), it is one of the largest teams assembled in the history of science. The head of the group is Fabiola Gianotti, an Italian woman who joined CERN in 1987. The Atlas group and another group working on the CMS detector will be at the forefront of the hunt for the Higgs particle at the

LHC. Both teams plan to share their data to maximize their chances of making a discovery. "It's been a long search for the Higgs," says Gianotti. "For me, it's a continuation of a long adventure. At some point, with this machine we are going to have the final word on it."

One intriguing possibility, after such a long race, is that in the lengthy shutdown at the end of 2012, scientists might combine results from the LHC with those from Fermilab's Tevatron. If both machines have seen hints of the Higgs boson by then, adding the data together might just be enough to claim a discovery. Both machines would end the hunt for the Higgs particle by crossing the finish line together.

There is far more to science than winning races. What physicists are really holding their breath for is news of the nature of the Higgs particle. Many physicists are hoping hard that, whenever the Higgs is found, it will prove to be a variety predicted by supersymmetry, the theory that requires a heavy and unseen partner for all of the particles we know of. A supersymmetric Higgs particle would open the door to a new realm of physics, one that takes scientists one big step closer to a theory that describes all of the particles and forces of nature.

The simplest possibility, which is enough to make it the most likely in some physicists' minds, is that scientists will discover a single uncharged Higgs particle that fits snugly into the Standard Model as the final piece of the jigsaw. That would solve the mystery of mass and tie up the only loose end left dangling from the Standard Model, but it would leave physics in a terrible position. Physicists know that the Standard Model doesn't describe everything in the universe. It says nothing about dark matter or the force of gravity. They desperately want the Higgs to show the way to understanding these and other mysteries.

Steven Weinberg at the University of Texas at Austin laments how particle physics, through no one's fault, has ended up in an unhealthy situation where theoretical ideas have got way ahead of what experimentalists can hope to observe in their experiments. The Higgs boson could be a way out of the problem, but this is not guaranteed. "Finding one neutral Higgs particle would not pull us out of the doldrums, it would put us into the doldrums," he said. "It would be just what we're expecting and it would give us no clue to anything new. Finding several kinds of Higgs, or even no Higgs at all, would be better."

The prospect of the hunt coming up empty-handed is a very real one, at least in some scientists' minds. In December 2009, only days after the LHC began its first tentative collisions, Tini Veltman, who shared the Nobel Prize with Gerardus 't Hooft for work involving the Higgs mechanism in the 1970s, gave a seminar at CERN. Veltman made the case that the Higgs simply might not exist and that particles acquire their masses in another way.[3] Another possibility is that if it does exist, it may be invisible, meaning that, once created, it decays into particles that modern detectors cannot see. Veltman concluded his talk by saying: "If no Higgs is seen in the near future, then there is either no Higgs or it's not visible. At this time, the 'no Higgs' possibility seems to be preferable to me."

Veltman is not the only one who wonders whether the Higgs particle exists in the form most scientists tend to think. In 2009, two physicists, David Stancato and John Terning, at the University of California at Davis, said the Higgs might not be a particle at all.[4] It could be what they call an "unparticle." Unparticles are a recent curiosity for physicists. They have a lot in common with regular particles, but come with a few quirks. Whereas normal particles have well-defined masses, the mass of an unparticle is smeared out and uncertain. That fact alone would make hunting for an Unhiggs problematic. To make matters worse, if an Unhiggs popped into existence inside the LHC, Stancato and Terning believe it would be almost impossible for detectors to spot it.

According to John Ellis, one of the main questions physicists need to answer when the Higgs boson is found is whether the particle is elementary or made up of smaller constituents. If it is made up from others, they could be particles we already know about or completely new ones. Finding out is likely to keep physicists busy for a long while.

Perhaps the most intriguing question surrounding the Higgs particle is best explained by an analogy John Ellis uses to describe its associated field. He asks people to imagine an infinite snowfield that fills the entire universe. Some particles can race over the snow as though they have skis on, as mentioned in the Prologue. These are photons, the massless particles that are oblivious to the Higgs field and so hurtle around at the speed of light. Other particles, like the electron, behave as though they have snowshoes on. They sink into the snow a little, make slower progress, and so acquire a small mass. And then there are particles like the top quark, which have no shoes at all. They sink deep into the snow and become superheavy in the process. The search for the Higgs boson is equivalent to gathering snowflakes from this cosmic snowfield and examining them under a microscope.

But why do some particles get bogged down in the Higgs field more than others? This question goes to the heart of the matter of mass. The heaviest kind of quark, the top quark, is roughly 400,000 times heavier than an electron. The Higgs field gives quarks and electrons a means of becoming massive, but it says nothing about why they have the masses they do. Why are some particles spectacularly more massive than others? "We have no idea," says Ellis. "We don't have a theory that explains why some particles wear snowshoes and others go barefoot."

The late British writer Ronald W. Clark tells the story of an unusual toast at the Cavendish laboratory at Cambridge University in his acclaimed biography *Einstein: The Life and Times*, published in 1972. The man in charge of the laboratory at the time was J. J. Thomson, who won the Nobel Prize for discovering the electron in 1897. The toast came from one of Thomson's staff, the eminent physicist Edward Andrade. He raised a glass to "the useless electron," adding: "And long may it remain so!"

No one knew the electron was going to transform the world. When Thomson gave talks on the newfound particle, people listened in disbelief. They thought he was pulling their leg: the idea of a particle smaller than an atom seemed absurd. Even those who believed it struggled to see how it might actually be of some use.

The story happens time and again in science. When James Clerk Maxwell unified electricity and magnetism in the late nineteenth century, he showed that light was an electromagnetic wave. Modern society could barely function without the technologies that exploit this knowledge. Manipulating electromagnetic waves gives us cellphones, wireless Internet, electronic car keys, hospital CT scanners, and satellite television. Again, no one saw the possibilities at the time. As Niels Bohr famously commented, prediction is difficult, especially when it's about the future.[5]

Some physicists have speculated that, just as we have learned to harness the electromagnetic field, so we will ultimately learn to manipulate the Higgs field. It's hard to imagine any obvious applications without plunging headlong into the folly Bohr warned us about. Altering the mass of fundamental particles isn't going to solve the world's obesity problem. Nor is it going to reduce traffic jams in New York, London, Tokyo, or Mexico City by enabling engineers to build vehicles that hover weightlessly above the ground. One scientist said, only half tongue-in-cheek, that the military could try to develop what might literally be called "weapons of mass destruction."[6] Turn the Higgs field off, and objects in that area would be gone in a flash, as their constituent particles suddenly lost mass and were able to move at the speed of light. This is unrealistic. It may be theoretically possible to alter the strength of the Higgs field, but it would take an impractical amount of energy.[7] Anyone bent on mass destruction could hold the Earth to ransom using far less energy in more conventional ways.

One more promising application of the Higgs particle is rapidly becoming one of the most exciting new areas in physics, and it is nothing short of profound. Since the beginning of the twenty-first century, a growing band of physicists has come to believe the Higgs particle could be at the heart of the next major revolution in science.

As upheavals in human thinking and perception of reality go, the revolution they have in mind is hard to beat.

Humans have always thought of themselves as special. As far as we know, we are alone in the universe in mastering language, turning out great works of art and literature, and formulating the laws of nature. But our view of humanity as the pinnacle of life has suffered huge blows at the hands of science. Every now and again, an idea comes along that is so revolutionary it rocks the foundations on which our hubris is built.

Three major revolutions have done more than most to knock humans off their pedestal. The first came from the Polish astronomer Nicolaus Copernicus in the sixteenth century. Copernicus used mathematical arguments to overturn the religious-minded view that the Earth was the center of the universe. Copernicus said it wasn't even at the center of our own solar system, but just one of many planets that orbit the sun.

We have Charles Darwin and his contemporaries to thank for the next revolution. More than three hundred years after Copernicus, Darwin provided a convincing case that humans—and every other form of life on the planet—were the result of a long process of natural selection. When he published his theory of evolution with *On the Origin of Species* in 1859, the implications were clear: humans weren't special; they were just another kind of animal.

The third revolution came nearly a century later when two Cambridge-based scientists, James Watson and Francis Crick, unraveled the structure of DNA in 1953. Their work marked the beginning of a new era of genetics and presented a further challenge to human dignity. To some scientists, we are but temporary guardians of selfish and self-replicating molecules, and our chief purpose in life is to pass them on to the next generation.

Today, physicists are seriously asking whether the Higgs particle is the cornerstone of the next major revolution in science, one that promises to be more profound than any that have gone before. The Higgs particle, they say, could be a bridge to hidden worlds.[8] If they are right, then what we call reality—that is, everything we know—is part of a much greater and more complex reality that we are completely oblivious to.

To James Wells, a physicist at CERN, the existence of hidden worlds seems almost inevitable once you concede that there is nothing privileged about human experience. He argues that, until recently, particle physics has been too anthropocentric. The particles that physicists have traditionally cared about are the ones that make up our bodies or that interact with those inside us. The point Wells makes is that there could be countless other particles and forces at work that we simply don't notice. "It would be really strange if everything that exists out there was stuff that our bodies feel. There is simply no reason why we should be so special," he says.

Scientists already know that the universe contains matter that we cannot see or feel. There are vast clumps of dark matter that lurk around galaxies that only reveal their presence by exerting a gravitational pull on the celestial objects around them. Cosmologists believe that dark matter makes up around a quarter of the mass of the universe.

How are these worlds hidden from us? Take a good look at the world around you. Everything, including this book in your hands and the chair you are sitting on, is made up from simple building blocks. Atomic nuclei are made from quarks that are bound together by the strong force. Around these central kernels are electrons that envelop the nuclei using the electromagnetic force. Atoms and molecules cling together to form everyday objects, thanks again to the electromagnetic force. The process repeats in countless permutations to give us the material world around us. All of these particles weigh something, because of the Higgs field and also the forces that glue them to each other. We feel their mass most often because the Earth's gravitational force pulls down on them. There is only one other force we know of, and that is the weak force that governs certain kinds of radioactive decay.

Why should the particles of matter we have found and the four forces of nature we are aware of be the only ones there are? There is no reason why the human body should be equipped to sense everything in the universe, and the existence of dark matter proves it is not. The extraordinary possibility is that there could be a host of particles and forces that are going about their business in a world that is entirely beyond our perception.

A hidden world might be made up of different kinds of dark particles and forces that are similar to those in our world but don't interact with the matter we know. There could even be a dark version of the Standard Model. If that is the case, what is to stop dark stars from being born and collapsing at the end of their lives or dark planets coalescing from dark cosmic detritus? Might there even be some form of dark life? "We don't have any evidence, but it's something we cannot rule out. There's no reason to think that a hidden world would be any less complex than the world we know. It could have just as rich a set of phenomena," says John March-Russell, a theoretical physicist who studies hidden-world theories at Oxford University. "Just as the Copernican revolution told us the Earth isn't special, the same could be true for everything that we've so far discovered. All of this stuff around us, the stuff of our reality, is it the dominant and most complex part of the universe? It might not be."

If hidden worlds are out there, the Higgs particle may be our best chance of finding them. Here's why. Some physicists believe that our own Higgs field will be tenuously linked to other Higgs fields that give mass to particles in the hidden world. This link forms a "bridge" that provides a way to peer into the hidden world and look at the kinds of particles from which it is made. The Higgs field is unique in being able to do this, for several reasons. One crucial feature that sets the Higgs field apart from all others is that it is a scalar field. This means that only the Higgs field looks the same from every direction. A second feature is that the Higgs field, and so its associated particle, permeates all of space. These two features make the Higgs field incredibly sensitive to tiny fluctuations in energy that might ripple over from the hidden world.

If the Higgs particle is found at the LHC, physicists believe that its death throes could shine a light into hidden worlds. The Higgs particle is highly unstable and is expected to decay into other particles almost as soon as it is created. If hidden worlds exist, the Higgs boson could decay into invisible hidden-world particles. These might then break down into "real" particles that we could see. The effect would show up in the LHC's detectors as a sudden burst of particle tracks that seem to come from nowhere. Spotting tracks like this would allow scientists, working backward, to build up an idea of the kinds of hidden-world particles the Higgs boson must have decayed into.

The idea of hidden worlds might sound as though it belongs to the ranks of crank ideas that physicists dream up in their spare time, but the prospect is being taken seriously. "If we can use the Higgs boson as a bridge to hidden worlds, it would be one of the most important insights we have had into space and time in the history of science," says Wells.

Peter Higgs retired from Edinburgh University in 1996, a full thirty years after making the visit to Princeton that gave such prominence to his theory of mass. He lives in an elegant Georgian townhouse beside a leafy park a stone's throw from the birthplace of James Clerk Maxwell, perhaps the greatest Scottish physicist who ever lived.

Higgs is often portrayed as a recluse, but like any convenient stereotype, the image is a flawed one. When the media sporadically turns its spotlight onto the hunt for the Higgs boson, the man behind the particle can be besieged with requests for interviews. They come from all over: radio stations in Moscow, TV stations in the Netherlands, and of course, newspapers across Britain.

Higgs is more elusive than reclusive. He prefers not to answer the phone unless he's expecting a call. He isn't on email and doesn't get involved with computers—a hangover, he says, from his student days, when it wasn't unheard of to spend all night waiting for a computer to churn out results. He is also busy in retirement. By the time he has had a chance to respond, most reporters' deadlines have long gone by.

After his seminal work in the 1960s, Higgs watched from the sidelines as experimentalists embarked on the hunt for the particle his theory highlighted. He has been dismayed along the way, but not by physicists' failure to make the discovery that would surely see him awarded the Nobel Prize. He is annoyed that some scientists have publicized colliders as Higgs-hunting machines and nothing more. Giant colliders look into the unknown. They might find evidence of deeply profound theories such as supersymmetry, extra dimensions of space, or something no one has yet even thought of. The Higgs is but one item on their list.

If Higgs were a betting man, he would put money on there being more than one type of Higgs particle. Ideally, he says, physicists will uncover a whole set of them that can be described by the mathematics of supersymmetry. So, nearly half a century after he first dreamed up the Higgs boson, is it important for him to see scientists discover the elusive particle? "It'll be a relief," he says. "It's been a long journey."

Epilogue

Shortly after seven o'clock on the morning of July 4, 2012, Rolf-Dieter Heuer, the director general of CERN, arrived at work and prepared to tell the world about a new particle in nature. The place was swarming with staff and students, and Heuer wanted to be sure the day ran smoothly. At the entrance to CERN, journalists and camera crews chatted in the sun as they waited for shuttle buses to the lab's main building. There, burly security guards who wore black heavy boots and occasional smiles kept order as the gathering swelled into a crowd.

Some who worked at the lab had been there all night. They came with laptops, duvets, and national flags, playing cards and alcohol, and spread out on the floor outside CERN's main auditorium. This was the venue for two morning talks on the Higgs boson, which Heuer was to introduce. To while away the hours before the doors opened, they talked and drank, slept and woke, and tapped half-heartedly at their computers. In the early hours, a more formal queue took shape, and with it, inevitable debates over the best route the human chain should follow. At one point, everyone stood up, shuffled along, and plunked themselves back down again. When the fire alarm went off at around 5:30 A.M., no one budged an inch, despite the best efforts of CERN's fire wardens.

When Heuer walked in, the queue clogged the stairs all the way down to the main floor from the auditorium's entrance on the floor above. Most of the people in line had no chance of getting in. There were already too many people upstairs. On a normal day, Heuer took the elevator to his office on the fifth, but this time he jabbed the button for the first floor. It was too crowded to walk up. Moments later, the doors opened on a throng of people, who broke into cheers as Heuer poked his head out. "Okay, okay!" he said above the racket, and with a wave continued up to his office. "I just wanted to see how it was looking on the first floor," Heuer said later. "It was a fantastic atmosphere."

The security guards let people in at 7:30 A.M. They counted groups of ten at a time to keep the right side of the room's capacity. The front half of the auditorium was reserved for CERN bigwigs and guests. That day, at least four of those to be seated in the front rows had once been director general, including Chris Llewellyn Smith, who in 1994 approved the decision to build the Large Hadron Collider. There, too, was Lyn Evans, who oversaw the construction of the huge machine. Behind him was John Ellis, who in 1976 told physicists how to find the Higgs boson amid the debris of colliding particles.

Others in the room could trace their role to the very beginning of the story. Two rows behind John Ellis sat the white beard and black-framed glasses of François Englert, who, with Robert Brout, wrote the first paper on what became the Higgs mechanism in June 1964. Brout, who had been ill for some time, died in May 2011, before the first hints of the particle appeared in the LHC. Across the aisle from Ellis sat Gerry Guralnik and Dick Hagen, who had flown in from the United States after hearing enough from physicists back home to convince them the journey was worthwhile. Their paper on the theory, with Tom Kibble at Imperial College, London, was published in October 1964. Kibble had a previous engagement that day and hadn't made it to CERN. Late into the room, and applauded like a baseball veteran returning to the mound, walked Peter Higgs, age eighty-three, who took a seat behind Guralnik and Hagen.

In official parlance, this was CERN's latest "Higgs search update," a humdrum title that promised little beyond the most incremental news of the lab's hunt for the missing particle. There was nothing special about the Independence Day timing, either, except that it coincided with the start of a major physics meeting, the International Conference on High Energy Physics (ICHEP) in Melbourne, where more detailed presentations were planned. That day, the heads of the two largest LHC experiments, Fabiola Gianotti on Atlas, and Joe Incandela on CMS, were down to speak at CERN in a session to be webcast live and beamed directly to the ICHEP audience already gathered in Australia. Since the seminar had been called on June 22, the rumor mill had gone crazy. Word had it that CERN had big news on its hands.

Minutes after nine o'clock, Heuer, a walking grayscale of charcoal suit, black shirt, and silver hair, took hold of a microphone from a CERN technician and turned to the audience in the auditorium. At the back of the room, students weary from camping overnight raised their heads and rubbed their eyes. Near the front, physicists who had spent their careers hunting the Higgs hung on every word. The spectacle was about to begin. Heuer welcomed them all and thanked the hundreds watching from Melbourne for joining in. With both hands cradling the microphone, he began. "Today," he said, "is a special day."

At this moment, both Gianotti and Incandela were in a curious situation. Sitting in the audience, and about to give the talks of their careers, neither in recent weeks had had what any reasonable person would consider adequate sleep. Their work had been grueling, meticulous, and stressful. Both knew, of course, what they were about to say and what they were about to show on the hundreds of slides they had prepared. But in the heave to get the work done, the enormity of their message had not quite yet sunk in. That would only happen later, when they would see with fresh eyes what those in the audience would soon see for themselves.

O ne year earlier, in July 2011, the telephone had rung at Peter Higgs's apartment in New Town, Edinburgh. The call was unexpected and went unanswered, because Higgs rarely answered unexpected calls. He was busy, too. His youngest son, Jonny, had moved in temporarily with his wife and two children while major work was being done on a flat they had just bought. The arrangement was pragmatic and loaded in favor of Higgs senior: he made them all breakfast, while his guests provided dinner. When Higgs got to his phone messages, he heard the voice of a neighbor, Jane Murray, who lived around the corner. As it happened, her son, Bill, was a physicist at CERN and worked on the Atlas team's search for the Higgs particle. From memory, Higgs recalls that the message went like this: "We've just heard from Bill. He asked me to tell you that something quite interesting is going to be announced at Grenoble next week." The reference to Grenoble meant the European Physical Society meeting, where both Atlas and CMS had talks scheduled for the twenty-second of the month. There, they would share their latest progress in the hunt for the missing particle.

It was clear before the Grenoble meeting that the Large Hadron Collider, and the Atlas and CMS teams, had outperformed expectations. The scientists had gathered as much data by July as they had hoped to compile by the end of the year. But the talks at Grenoble were equivocal. Atlas and CMS ruled out much of the mass range where the Higgs boson might lie. The intriguing feature that sparked Jane Murray's call appeared in the space left over, a gap between 120 GeV and 140 GeV. Both groups saw small bumps in their data, perhaps from extra particles produced as Higgs bosons decayed. Looked at from the right angle, and with a positive frame of mind, these might just be the first glimpses of the missing particle. "There was a hint of excess data, but it was no more than a hint," Higgs said later.

The LHC pushed on through the autumn, smashing particles and churning out data. By the end of the 2011 run, at 5:15 P.M. on October 30, the machine had racked up 400 trillion proton-proton collisions per experiment. This surpassed the goal for the year five times over. The collider had produced data at 4 million times the rate of its first run in 2010. If Higgs particles were for real, then scores of them had already popped into existence and disintegrated in the LHC in ways that might be detected. Each one would have existed for the briefest of moments before decaying into more common particles.

As Atlas and CMS crunched the 2011 data, CERN arranged back-to-back seminars for the teams to share their results. On December 13, lab staff gathered in the main auditorium to hear Fabiola Gianotti on Atlas, and Guido Tonelli, then head of CMS, present their findings. Their stories were in remarkable agreement. Both ruled out a Higgs particle across the whole mass range, save for one intriguing sliver, around 125 GeV. Here, the experiments saw the strongest hints yet of a Higgs boson. Each team had a bump in their data, with a statistical significance of about two standard deviations, or *sigma*. While some physicists tangled over how optimistically to interpret the numbers, Gianotti stuck to the facts. "We see some interesting signals," she said. "But we need more data."

For those working on the LHC, each year closed with a winter shutdown, when engineers stepped in to do essential maintenance on the collider. Before it was set running again, the CERN accelerator team headed for Chamonix in the French Alps to thrash out how to get the most from the machine and maximize its chances of helping them make a discovery. More often than not in particle physics, that meant more collisions, or higher energy collisions. Or better still, both.

In 2012, the LHC accelerator physicists pushed for both. The decision was always a balancing act. Squeeze too little from the machine, and discoveries were harder to make. But push the collider too hard, and you risked catastrophe, like the short circuit in 2008 that blew a hole in the machine and shut it down for more than a year. The decision was particularly acute in 2012. CERN was under great pressure to make a discovery before year's end, even if that meant ruling out the Higgs boson. That winter, the collider was to close for two long years to allow servicemen to descend inside the tunnel for extensive work—beefing up its components so the collider could begin running at its full design energy in 2015 or thereabouts. To shut down empty-handed would leave physicists with nothing new with which to grapple in that bleak down period. That was not the only problem. Many of CERN's member states in Europe at the time were seeing either financial turmoil or changes in government, and quite often both. Had the LHC found nothing, those countries might cut their funding to CERN. But the lab could not afford another catastrophe. Steve Myers, a straight-talking Belfast man, had taken over as director of accelerators on September 18, 2008, only fifteen hours before the accident occurred that tore a hole in the LHC. He summed it up like this: "If we had blown another hole in this machine we would have lost our reputation forever."

The engineers hatched a plan at the end of the retreat in February. Myers's team would nudge up the energy of the LHC's beams so that the hurtling particles would meet in head-on collisions with a total energy of 8 TeV. Through 2010 and 2011, the machine ran with 7 TeV collisions. The move pushed the machine a little harder, but Myers calculated that the extra risk was marginal, with the chance of another damaging incident still around 1 in 100,000 for the year. In packing more punch, the machine should make roughly 30 percent more Higgs particles than before. That was one change they agreed on. Another was to squeeze the LHC beams at the points where they crossed inside the detectors. This would force the particles into a more confined space, so when bunches of particles met, more of them would slam into one another. Squeezing the beams from around 50 microns across to less than 19 would double the number of collisions per second.

The tweaks are easier to describe than they were to execute. When you fiddle with the LHC beam, an untold number of effects can come into play that threaten to ruin your day. Squeeze the beam at one point, and it flares out elsewhere, like a sausage-shaped balloon grasped firmly in the middle. To protect the machine from bulging particle streams, collimator jaws close in and absorb those that stray too far. But bring the collimators in too much, and lose just one-millionth of the protons, and the particles can scatter onto the machine, heat parts up, and cause the beams to collapse. Another problem: the circulating protons generate fields in their surroundings that the beams feel as *impedance*, or opposition. The smaller the gap between the beam and the collimators, the greater the impedance. And when the impedance gets too high, the beam loses stability and drops out. One more: protons lap the LHC around 11,000 times a second. As they circulate, they wobble in two planes: up and down, and left and right. If the number of wobbles per lap is a whole number, or, say, one-half, or one-third—in which case the wobble completes its movement over two or three laps, respectively—the oscillations build over time. Fail to control these "nonlinear resonances," and the beams will fail, too. Finesse is a critical art for an LHC engineer.

With so much to go wrong, it is a miracle all went right. Well, almost all. There was one heart-stopping incident in April, at the start of the LHC's 2012 run, that Myers will not forget in a hurry. Before particles are fed into the collider, they are whipped up to speed inside the aged accelerator at CERN called the Super Proton Synchrotron, or SPS. The moment the particles pass from the SPS to the LHC is exquisitely choreographed. As the particles enter the LHC, fast kicker magnets switch on that create powerful, instantaneous fields that steer them onto the right course. These magnets must kick in to deflect the particles. At this moment in time, the proton beam is moving at close to the speed of light and carries around 2 megajoules (MJ) of energy. That is enough to burn a 1-meter hole in most metals. The nerve-jangling incident happened at 4:47 A.M. on April 15. Just as 108 bunches of protons passed into the LHC, a kicker magnet sparked and its field crashed to zero. Only 32 bunches made it into the LHC on the right course. The rest did not. Undeflected, they continued in a straight line. Had the risk not been foreseen, the beam could have punched a hole in the collider. But protection was in place, in the form of an absorber plate, positioned right in the path of the errant beam. "It blasted onto the protection device and luckily it was okay," Myers said later. "We have protection for it, but it's something you never want to test. For us it was a case of, thank God the safety mechanism worked. When it happens, you see it instantly. There are alarms going off everywhere."

The upgrades agreed in Chamonix led to an extraordinary run at the LHC in 2012. Shortly before 1 A.M. on April 5, stable beams of protons heading in opposite directions were crossed in the machine, colliding particles at 8 TeV, an unprecedented energy in a manmade experiment. In the three months from April to June, the machine generated more data than in the whole of the previous year. The success was due to the immensely capable accelerator team and the incredible control they had over the beams, particularly through feedback systems that fine-tuned the machine in real time.

There is a faux machismo in the accelerator team that is in keeping with the group's status as wielder of the world's largest instrument. As the team drove the machine hard in 2012, the LHC threatened to fill the Atlas and CMS detectors with more collisions than they could process. Had the problem, known as pile-up, become insurmountable, the accelerator team was ready to ease up on the intensity, but they hoped that Atlas and CMS could rise to the challenge. As Mike Lamont, head of accelerator operations at CERN, told the journal *Nature*, at the time: "If they can take it, we'll give it to them." In practice, that meant the detector groups had their work cut out for them to deal with the surge of collisions the machine was now delivering.

Gianotti and Incandela saw it coming. The beams in the LHC carried 1,380 bunches of protons each. The intensity of the beams was so high that every time 2 bunches of protons met head-on in the machine, they got 30 to 40 particle collisions. Most of the time, the collisions did not produce Higgs particles, but showers of more familiar subatomic fragments. These left tracks in the detectors and became a huge drain on the computing power needed to sift the chaff from any valuable traces of Higgs bosons. At CMS, Incandela worked out that unless they did things differently, they would have to raid the memory from a third of their computer chips just to process the data from the collisions—and the time required to process each event had nearly doubled. That left them with only two-thirds their usual amount of raw computing power to do almost twice as much work. Incandela launched a task force to work on the problem in late 2011, and by February, through ingenious tweaks to the CMS algorithms, including the "triggers" that record only the most promising-looking collisions, the team effectively overcame the pile-up problem and even significantly improved their ability to extract Higgs signals from the data. The Atlas team made similar improvements. At the start of the 2012 run, both experiments were as ready as they could be to take all the data the LHC could provide.

The two teams analyzed their first haul of 2012 collisions a month or so after the run began. At this point, they were not looking for fresh signs of the Higgs boson. Under strict instructions from Gianotti and Incandela, they looked only where the Higgs particle had already been ruled out and in event samples where the signal was not expected to show up. The narrow window of masses where hints of the Higgs cropped up in December was firmly off-limits. The restriction, known as "blinding," might sound harsh and unnecessary, but it was absolutely crucial. The process ensured that the scientists would not see things early on in the data that were not really there, developing biases that would skew the analysis down the line. The procedure also allowed the teams to ground-truth their analyses by checking that the background signals the detectors picked up from well-known particles matched their expectations. When they were happy with their analyses, they could unblind the rest of the data and see what was there simply by tapping a few keys. "To unblind you just take out the lines of code that say 'do not plot here,' rerun and let it rip," Incandela said later.

From raw collision to dot on a plot takes a chain of events that draws on hundreds of people with expert knowledge. During LHC runs, collisions were recorded at the CMS and Atlas control rooms twenty-four hours a day, seven days a week. On Atlas, for example, a crew of nine on a rotating shift ensured that data were collected as efficiently as possible. They reacted to problems immediately. On call was another layer of experts who helped out as needed. Other groups met every morning to make sure the detectors started the day in full working order. This system was pivotal in enabling the detectors to make use of an impressive 90 percent or more of the data the LHC generated in 2012.

The freshly recorded data were sent directly to CERN's computing center. Here, around 10 percent went into an express data stream that was processed immediately, with people checking calibrations and ensuring that particle tracks in the detectors were properly reconstructed. Once that data had the "all clear," the rest was bulk processed and sent over CERN's powerful grid network to worldwide computing centers, where remote teams got to work.

The Higgs boson can decay in several ways, and each leaves a specific signature in the detectors. For example, some of the time, the Higgs particle will disintegrate in a flash of two high-energy photons. At other times, it will vanish in a spray of four leptons, such as electrons—or muons, their heavy relatives. Several groups worked in parallel on each signature, or channel, and counted independently the number of candidate Higgs events they saw. These teams fed into working groups that looked at all the results for each channel. Finally, the channels were combined to give an overall result for the detector, which then went out to the whole collaboration for approval.

Unblinding is the moment of truth. Up until that time, all outcomes were possible. The particle might not exist, and be ruled out beyond doubt. It might be there in the data and pop up clear as day. Or the results might be inconclusive, leaving scientists none the wiser. In December 2011, Rolf-Dieter Heuer, the director general, had said he would have an answer by the end of 2012. But since then, Myers's team had pushed the LHC harder, and both Atlas and CMS had upped their game. The teams unblinded their data in June. From that moment on, the world never looked the same again.

A month earlier, Peter Higgs was in Bristol. He had traveled down to take part in a public discussion at the city's Festival of Ideas as well as to do some filming for the London Science Museum around his childhood home and at Cotham School, which he had not seen since leaving in 1946. On May 17, he gave a talk at Bristol University on the history of the famous boson. Afterward, he fell into conversation with physicists who worked on CMS. Having just lectured on the particle's past, he now learned about the future of the missing boson. Much more work was needed, he heard. The particle was not about to be found.

Atlas and CMS stopped taking data at 6 P.M. on June 18. By that time, both teams had already unblinded their analyses. A week or so earlier, Gianotti had set eyes on an Atlas plot that showed what might be Higgs particles decaying into photons. There was a huge spike in the data. The bump was too big to be the Higgs boson, but as more data came in, the signal shrank, eventually stabilizing at a level more reasonable for a Higgs particle with a mass of around 126 GeV. In mid-June, Atlas unblinded the other main channel the physicists were watching, where Higgs particles decayed into four leptons. There was almost nothing there. But as more collisions were fed into the analysis, a bump in the data began to grow.

On June 14, at around 10:30 P.M., Incandela was on the phone to Chiara Mariotti, a colleague on CMS. She was working on a channel called Higgs to ZZ, which measured the decay of Higgs particles through two Z bosons into four leptons. Incandela hadn't called about this—the CMS team had set aside the next day to unblind their data—but Mariotti was struggling to contain herself. "She had just seen the plot. She was so excited she sounded like she was on laughing gas," Incandela recalled a few weeks later. He put on an Italian accent to repeat what she said: "It's beautiful! It's beautiful!" He told her: "Okay, you've got to send me this." It was too risky to upload the plot to the web. Instead, Mariotti took a screenshot and beamed it over.

The plot that arrived at Incandela's house that night had a perfect bump at around 125 GeV. It was a textbook signature for the Higgs boson. "I realized instantly my life had changed forever," he said. "When I first saw the plot from Chiara I had a few minutes of joy followed by an unbelievable avalanche of worry. I knew where it was heading, what had to be done. We had to verify everything, we had to look at all the other channels, we had documentation to complete, and we had to worry about keeping it all quiet until we'd checked everything. As the head of the experiment, you don't really have the opportunity to sit back and enjoy it. I couldn't sleep that night. Lists were forming in my head."

Peter Higgs had no idea what was happening at CERN. He had spoken to physicists on CMS in Bristol weeks before the strength of their results became clear. No wonder he left thinking there would be no news soon. The same story was about to play out again. On June 19, Higgs traveled to Cambridge University to receive an honorary doctorate. He spent a couple of days in the city, and while he was there, caught up with researchers in Gianotti's Atlas group. Again, he got the impression that discovery was not at hand. The signals, he heard, were not strong enough to reach the magic 5-sigma level required to claim a discovery. Higgs went back to Edinburgh blissfully unaware of the tsunami that was coming.

When Higgs got home, there was a telephone message from a Dutch filmmaker, Jan van den Berg. Van den Berg and his colleague, Hannie van den Bergh, had made an award-winning documentary a few years earlier called *Higgs: Into the Heart of Imagination*. They had always hoped to update the documentary with an epilogue and broadcast it the day the particle was discovered. Van den Berg wanted to get footage of Higgs in Edinburgh in anticipation of the particle being found later that year. The holiday season was close, and film crews would be hard to find if they waited much longer. Higgs was happy to take part, but told the filmmaker he was busy that weekend. "By the way," Van den Berg recalls him saying. "I'm told that the CERN analysts definitely need more time, until September or so." The two men agreed to meet in Edinburgh in late July.

The day after Higgs and Van den Berg spoke, Heuer held a meeting with Incandela and Gianotti. He wanted to know where their analyses stood. Both said they were above 4 sigma. The results were enough to press home a new problem. The ICHEP meeting in Melbourne would begin with parallel sessions, with scientists on Atlas and CMS presenting results from each channel of the Higgs search separately. The plenary talks that summarized those results would come only at the end of the meeting. It meant the news would come out in fragments. By this time, the CERN council had already asserted that the results, whatever they might be, should be given first at the lab. And so, on Friday, June 22, CERN announced that it would hold two seminars and a press conference on July 4, at which both Atlas and CMS would unveil their news on the Higgs for the first time.

The date for the seminars gave Atlas and CMS less than two weeks to finalize their results. The last collisions had to go through the hands of hundreds of people to be elevated from raw to approved data. On the evening of Sunday, June 24, Gianotti was at home, looking at data and checking results, when two plots arrived from her collaborators. The graphs included all the Atlas data for the two most important channels, namely, Higgs particles decaying into either two photons or four leptons. The Higgs signal had grown to around 5 sigma. Her reaction to seeing the plots was something more complex than pure excitement. A few weeks later, in language infused with the terminology of a particle physicist, she said: "There were big spikes of emotion and excitement to look at the plots; those peaks were the sign of a Higgs-like particle and at the same time the result of a huge amount of painstaking work, intensity, and dedication by many people. I don't know which of the two things was making me more proud. Those weeks were something I had never experienced in my life before. It was very intense, very stressful. We had very little time to make sure we made no mistakes, but we were so happy, the emotion was palpable."

The following morning, Jan van den Berg walked into Stan Bentvelsen's office at the National Institute for Subatomic Physics in Amsterdam. Bentvelsen was on the Atlas team and happened to be the main character in the Dutch documentary. Van den Berg only dropped by to catch up, but his timing was perfect. Bentvelsen had just seen the latest Atlas plots. "I found Stan with a huge smile on his face," Van den Berg recalled. "He said something like: 'Just look at this. From now, I cannot deny it anymore. This very much looks like what we've been looking for so long.'" A few hours later, Van den Berg was making frantic plans for what he called a "hit and run." First, he had to fly to CERN with Bentvelsen to film the Dutch Atlas group preparing for their seminar. Then he had to film Higgs, who had just flown out to Sicily for a summer school in Erice, an ancient hilltop town on the island's western tip. Van den Berg got a call through to Higgs the next morning. He said he was coming to film in Sicily that weekend. "We've been seeing things we were not supposed to see, and hearing things we were not supposed to hear," he told Higgs.

The filmmakers flew to CERN with Bentvelsen on June 28. Jan van den Berg and Hannie van den Bergh spent two days filming his team as well as other CERN scientists who had appeared in the original documentary. They included Rolf-Dieter Heuer, John Ellis, and Sergio Bertolucci, the lab's director of research. They told Van den Berg that they couldn't tell him anything. But their grins said it all. "I had never seen that kind of excitement among physicists before," Van den Berg recalled. "All over CERN there were people wearing big smiles."

That Friday, Heuer held a final meeting in his office with Gianotti and Incandela. This time, both teams had Higgs signals of around 5 sigma. The analysis was so hot, so fresh from the teams, that neither Gianotti nor Incandela had fully absorbed the results. Heuer felt a similar dissociation from the figures before him. Speaking in his office a few weeks later, he fished for words to explain, found a handful, then said nothing more for five or ten seconds, instead beating out a rhythm on his desk, tap, tap, tap, with a half empty bottle of water. "Were we excited? That's the funny thing. We saw we had a discovery, but we were not yet in a state of realizing what that meant. We weren't jumping up and down. We hadn't digested it then."

That evening, the filmmakers and Bentvelsen boarded a plane from Geneva to Rome, and from there, flew on to Palermo. They arrived around midnight, drank some beer, and grabbed a few hours of sleep. Early the next day, they drove the 50 kilometers or so to Erice. Van den Berg had sketched out the scenes he wanted to film. Bentvelsen and Higgs would bump into each other on a quaint backstreet in Erice, and, for the camera, act as though they had never met. In fact they had met before this, but only briefly; it had been at a showing of Van den Berg's documentary in 2009. The two would then take a walk to a stone bench with a breathtaking view of the bay, where Bentvelsen would pull out his laptop and show Higgs the latest Atlas plots. After a quick toast over a glass of wine, Bentvelsen would wrap up the shoot by telling Higgs he must come to CERN for the discovery announcement.

Filming the scenes was not as simple as scripting them. Erice is a tiny place and was swarming with physicists at the summer school. This posed a problem for Bentvelsen. He was worried he might be seen with Peter Higgs and a film crew, especially as Fabiola Gianotti had just sent out an email forbidding anyone on Atlas from disclosing unapproved results outside the collaboration. The filmmakers had protested. They were not journalists, but artists. The film was an artistic impression of the things going on in physics. A few weeks later, recalling that day's filming, Higgs said, "Stan was feeling guilty about being there because some of his colleagues on Atlas were teaching at the summer school and might think he was leaking things which should be confidential," he said. "Which in a way he was."

The filmmakers pressed on. With the first scene, they got off to a shaky start. Higgs and Bentvelsen bumped into each other, as planned, but quickly drifted off-script by talking about old times. As the two walked down the cobbled street, the conversation flowed as naturally as it can between two physicists who barely know each other, speak different languages, and have no history of acting. "At the end of the day, we've been looking for the Higgs particle for so long," Bentvelsen tells Higgs in one scene. Higgs nods. "For me it's been forty-eight years," Higgs replies. "It may be just in time, no?" says Bentvelsen. The line leaves his octogenarian companion silent.

The bench scene ran into its own difficulties. Bentvelsen fired up his laptop and began to show Higgs some recent Atlas plots. But as he shared them, Bentvelsen kept saying they were old results, from the previous December. Van den Berg was dismayed. "I was constantly worried about how to edit those sentences out. In every sentence he was saying the wrong things. I just wanted him to say 'I think we've got it!'" Higgs, meanwhile, was struggling with what the plots meant. His role in the scene was to look increasingly happy as the results sank in. "We got to the bench, and Stan showed me some plots, but frankly, I don't know how to interpret them. People have to tell me that's the bump that's significant, so it took several takes for me to look happy enough," he said.

At the end of the shoot, the group headed for lunch at the nearby Venus restaurant. As they ate and talked, Alan Walker, who had traveled with Higgs from Edinburgh University, took a call. It was John Ellis at CERN with a message for Higgs. No one can recall the exact words, but the message Higgs got was that if he didn't get to CERN for the seminars, he would probably regret it. Rumors in the United States had already convinced Dick Hagen and Gerry Guralnik to fly over for the talks, and Ellis thought Higgs and Tom Kibble should know. Separately, James Gillies, head of press at CERN, contacted Higgs, Kibble, and François Englert in Brussels. They were welcomed along, but not formally invited, because it was still unclear what CERN would say. "I don't think it would have been very kind if we'd invited them for a stupid update, so we left it up to them," Heuer said later.

The call threw lunch at the Venus into something close to chaos. Higgs and Walker had to change their travel plans and get to CERN. There were flights to cancel and others to book, and expiring travel insurance to ignore. The two men quizzed each other about their stocks of underpants and happily concluded both had enough clothes for the diversion to Geneva. The waiters at the restaurant could only look on, bemused by the sudden uproar at the table.

Ever since the LHC had gotten its stride, in 2010, the machine had eclipsed the Tevatron collider near Chicago. But as aged and destined for recycling as it was, the Tevatron kept up the search for the Higgs to the end. Even when it closed down, in September 2011, after the U.S. Department of Energy refused to fund an extended run, the machine had stayed in the game. On July 2, 2012, the Higgs-hunting teams on the Tevatron's CDF and DZero experiments announced their final word on the missing particle. Their combined results showed a broad bump in the data, perhaps from a Higgs particle with a mass of somewhere between 115 and 135 GeV. The strength of the result stood at 2.9 sigma, far from enough to claim a discovery. The day the Tevatron delivered its parting shot, Higgs arrived at CERN with Walker. They got in late and headed to the canteen for dinner. Before long, a queue formed for Higgs, as old friends and former students from Edinburgh spied him and began to stop by for photos and autographs.

On Atlas and CMS, the run-up to Wednesday, July 4, became ever more frantic. The day before, Incandela had fifty pages of notes and hundreds of slides to sort out. Many had to be reformatted for the seminar the next day. To work through the material, he set up a situation room down the corridor from his office and filled it with bright youngsters who were experts in the analyses. They were in constant touch with the rest of their teams. He was sure some of them looked thinner than they had six months ago.

That afternoon, as they plowed through the slides, Bob Cousins of the University of California at Los Angeles, one of Incandela's team members, stopped typing. "Uh oh. Uh oh. Uh oh," he said.

"What is it?" asked Incandela. "You don't want to know," came the reply. But there was no point in hiding it. Incandela had just recorded a video for CERN in which he discussed, albeit vaguely, the team's results. The interview had been due to go live, alongside a similar video from Gianotti, after the seminars the next day. Somehow it had found its way onto the Internet. The story is that Incandela had been unhappy with an earlier version (he'd used the "d" word, meaning "discovery") and recorded a replacement interview in which he used "observation" instead. When the CERN video team uploaded the file, through glitch or error, it lost its security settings and immediately went public. Blogs had already picked it up. CERN, the organization that had invented the Internet, and given it free to the world, had leaked its biggest story in recent history by screwing up its web security.

"It was everywhere," Incandela said some weeks later. "I was getting messages from friends in New York about it. I couldn't believe it. I said, look, I don't have time to even think about this. It's embarrassing, but there is nothing I can do. I'd reached the point where nothing else could bother me. You're already at such a high level of stress, it doesn't matter." Gillies in the press office had some words of comfort. The video had given out far less information than had already leaked from CERN in less overt ways and made its way into blogs and news reports. The CMS team pushed on, working through the slides for the next morning's seminar. The pressure was extreme. "I knew it was going to be very, very close and so did Fabiola [Gianotti]. I remember a ten-day period when several times every day I had to tell myself not to panic. I just didn't see how we'd get it all done. I didn't see how we would make it. And then, on the night before, for the first time I thought, okay, I think we're going to make it," Incandela said.

On the other side of the building, Gianotti was finalizing slides ahead of her own seminar. She completed them at around one or two o'clock in the morning and slept like a log. Incandela still had slides to finish, but was close to complete exhaustion. He went home and grabbed five hours of sleep, the most he'd had in weeks. He was due to speak first the next morning, at nine o'clock.

Peter Higgs had a calmer day. Gillies welcomed him to his office, declaring it the only part of CERN that was not leaking. Someone handed Higgs the press release CERN planned to put out the next day. He had a quick read-through, and then recorded a video interview, which CERN managed not to leak. In the press release, both Gianotti and Incandela talked of 5-sigma signals, but never mentioned the word "discovery." That word appeared only once, in a quote from Rolf-Dieter Heuer. He said physicists had reached a "milestone" in their understanding of nature. "The discovery of a particle consistent with the Higgs boson opens the way to more detailed studies . . . and is likely to shed light on other mysteries of our universe," one quote read. Higgs got to see the release early so he could give his reaction in the video. "I was supposed, with the aid of that, to look happy," Higgs said later that month. "I think I managed. It was the first official indication I had."

The wording of the press release might seem simple, but it was arrived at after days of debate. The discussion, gleaned from interviews and email exchanges with Incandela, Gianotti, Heuer, and Fabio Zwirner, chairman of the CERN Scientific Policy Committee, who was consulted on the wording, reveals much about the psychology of the physicists involved at that time and the way particle physics is done in the modern era.

The Atlas and CMS groups were uncomfortable using the word "discovery" for several reasons. In particle physics, a result with a statistical significance of 5 sigma is considered strong enough to count as an "observation," and equivalent in most situations to a formal discovery. In real terms, the odds of a 5-sigma result happening by fluke is around 1 in 3 million. So why not come out and call it a discovery? One reason is that "discovery" has connotations—of reward, for example—that somehow aren't in keeping with the purest scientific ideals. The word "observation" is neutral: an objective statement of fact. It is simply the preferred term. But issues of process play a role, too. Prudent scientists tend to claim official discoveries only once their papers are published, that is, after the results have been fully scrutinized, peer reviewed, and given the thumbs up. At the time of the seminars, the results were preliminary and not ready for publication. Another factor was that Atlas and CMS had only just reached the magic 5-sigma level, or come within a whisker of it. The analyses were done at breakneck speed, and the teams needed time to trust them. Lurking somewhere in the background was an awareness of the cost of making a mistake in full view of the world's media. In an email some weeks later, Zwirner expressed his personal view on this: "There was the feeling that a mistake, which is always possible in forefront science, could give a fatal blow to the credibility of CERN, of the LHC experiments, and of particle physics at large, and with the recent memory of the case of superluminal neutrinos everybody wanted to be ultra-cautious." Evidence for supposedly superluminal, or faster-than-light, neutrinos had been unveiled at CERN in September 2011 by scientists on an experiment called OPERA based at the Gran Sasso Laboratory in Italy. As many physicists suspected, the measurement was faulty, but that's another story.

The final wording for the CERN press release was nailed down when Incandela and Gianotti met with Heuer and Sergio Bertolucci. They agreed that neither team would use the word "discovery," but decided that Heuer could, on the strength of their combined results. It made sense: whatever the phrasing they agreed, history was sure to remember the seminars as the discovery talks. "They were not willing enough to say they had a discovery, because they were very cautious, and that was fine," Heuer said later in his office. "But me seeing both results, I could speak about a discovery. It's my job to stick my neck out."

The night before the July 4 event, Peter Higgs enjoyed a quiet celebration. John Ellis had invited him, Alan Walker, and Chris Llewellyn Smith to his house in Tannay, a small village a short drive north of Geneva, where CERN is based. The house sat in an old apple orchard and was shrouded in an out-of-control wisteria. As John's

Colombian wife served up *pan de yuca* and got to work on the filet mignon and a soufflé, made with apple and mint from the garden, the physicists went out into the warm summer evening and cracked open a bottle of champagne. (At their request, Ellis later donated the empty bottle of Perrier-Jouët to London's Science Museum.)

First through the doors of the auditorium on the morning of July 4 were those who had camped overnight. Some headed to the back, where it wasn't unheard of to close your eyes and rest your head on the bench in front. Senior figures at the lab, at least the ones with reserved seats, filed in later, taking up the front rows. The room got noisier and noisier: a home crowd before the sure win that would seal their dominance this season.

Over at building 40, Incandela was still working on his presentation. He finished at 8:42 A.M., uploaded the file a few minutes later, and walked over to the auditorium. Shortly after nine o'clock, Heuer was up in front of the audience. He introduced the two speakers and invited Incandela to take the floor. There was a lot to say. Incandela had compressed the work of probably thousands of people over the years into about a hundred slides, and had another seventy or so as backup to handle any questions that arose. One by one, he took the audience through them. He praised the LHC's performance and described how CMS wrestled with pile-up. The first result that mattered was evidence for Higgs particles decaying into two photons. The slide showed a red line through black dots that traced a beautiful, clear bump in the data. At this point Incandela stopped talking. He stood back and simply stared at the slide. After five, maybe ten seconds, he flicked on to the next slide. "I was lost for a moment," he said. "Excuse me."

The results built up slowly. Taken together, the cleanest channels, in which Higgs particles decayed into either two photons or four leptons, gave a solid 5-sigma signal. When the slide went up that stated as much, the auditorium broke into cheers and applause. Incandela picked out Steve Myers in the crowd and walked over to shake his hand. "It was the last month of running that did it," Incandela said to the room.

Incandela had plots from three other channels, and some showed more curious results. The measurements of Higgs particles decaying into two tau particles, those heavy cousins of the electron, revealed no sign of the Higgs boson at all. The result, perhaps a statistical effect, perhaps the first glimpse of new physics, pulled the combined signal down. Taking all the channels together, CMS reported a 4.9-sigma signal for a Higgs boson with a mass of around 125 GeV.

A few weeks later, Incandela recalled how the significance of the results really sank in as he spoke. "I realized that everything we had to do was done, from the night Chiara [Mariotti] sent me that plot, up until the talk, which I had just finished. We'd made it. I remember giving the talk and at a certain point I showed this plot with a bump and people in the audience gasped. I stood back and thought I'm just going to linger on this for a few seconds. It really hit me then: we've really discovered something. I started to enjoy it during the talk. That was the first moment I'd had to relax, and it was the most high pressure talk I'd ever given in my life."

Fabiola Gianotti spoke next. Her talk pulled together analyses of the two highest resolution channels, namely, Higgs particles decaying into two photons or four leptons. Individually, both results were strong. But when Gianotti clicked on the slide that showed a combination of both results, the room erupted. The trigger was highlighted in a little red box. It said "5.0 sigma." The ovation was rapturous. Gianotti tried to calm them down. "I'm not done yet, there's more to come, be patient," she told the crowd, but they had heard all they needed to hear. The Atlas group's results pointed to a Higgs particle with a mass of 126.5 GeV. "I didn't realize it was going to be a historic moment," she said in her office a few weeks later. "We were so concentrated on completing the results, and making sure we presented them in a solid way. I didn't realize until I was there."

When Gianotti wrapped up, the audience broke into thunderous applause once more. They stood and cheered and stamped and whistled. This was the payoff for more than twenty years of work at CERN, from understanding how the boson might rear its head in colliders, to building the LHC and hunting the particle down. The crowd's reaction was too much for Peter Higgs. He pulled a crumpled white handkerchief from his pocket, removed his glasses, and wiped tears from his eyes. He clapped, slower than the rest, and seemed a man in shock. He looked up and around, as if at a loss of what to make of it all. At the front of the room, Heuer grabbed a microphone and turned to the audience. "As a layman, I would now say I think we have it. You agree?" The answer came as a roar. "We have a discovery. We should state it. We have a discovery," Heuer said. Soon after, a text message appeared on his cellphone from a friend at the German DESY accelerator. It said: "Now enjoy the champagne on the plane."

The microphone went next to the four attending theorists who had started it all in 1964. Higgs, with jacket collar sticking up but composure regained, congratulated the CERN staff on their achievement. "For me it's really an incredible thing that it's happened in my lifetime," he said. Englert had come down the stairs to stand next to him. The two men had never met before this day. Englert praised the CERN staff, too, and regretted only that his lifelong friend and collaborator, Robert Brout, had not survived to witness the day. A few steps down, Guralnik was impressed at a physics event where the applause matched that of a football game. Hagen was "overwhelmed, by the quality of the machine, the quality of the analysis, and the intense labor which has gone into getting this result."

Back in Amsterdam, the Dutch filmmakers, Jan van den Berg and Hannie van den Bergh, had watched the seminars live on the web. They grabbed some last-minute scenes for their documentary—of Heuer announcing the discovery, and Higgs's reaction to the news—and went on air that night after the evening news. The scenes of Higgs in tears were moving for Van den Berg. "It was a combination of my affection and admiration for Peter, the fact that we had met only a few days before, filmed together and shared a few glasses of wine, and sheer exhaustion, without doubt," he said.

Earlier that morning, the burly men in black heavy boots had looked out of place among this crowd. But their presence was about to make sense. At the press conference after the talks, Higgs was mobbed by journalists and camera crews. The security staff kept them at bay as best they could, shielding an eighty-three-year-old man who, for that day at least, was a rock star. After a handful of interviews, Higgs and Walker made off for the airport, where they had seats on a cheap flight back to Edinburgh. On the plane, Walker asked if Higgs cared to celebrate with a bottle of Prosecco. Higgs declined but ordered a can of London Pride. The empty can might have made a nice memento, had Higgs not tossed it into the rubbish.

Incandela and Heuer left Geneva soon after. They had to get to Melbourne for the ICHEP conference. Incandela slept on the first flight, all the way to Singapore. The two men arrived in the middle of the night and made their way to one of the lounges. There, Incandela picked up a copy of the *Wall Street Journal Asia*. The splash was accompanied by a beautiful CMS collision event. "Wow, that was fast," he thought. On the connecting flight, Heuer grabbed an *International Herald Tribune*. It quoted both of them on the front page. The next thing Incandela knew, the flight attendants were treating him fantastically well. Heuer had waved the paper at them and showed them where Incandela was sitting.

After the seminars, Gianotti answered questions at the press conference, gave more interviews, and went from there to her regular Wednesday afternoon meeting with the LHC machine committee. She missed the drinks in building 40 where Atlas has its offices. Late that day, she left work, packed for ten days in Melbourne, and went early to bed. In the morning, she climbed on the plane and slept for hours.

Wednesday, July 4, 2012, will forever be discovery day at CERN, but on that day, at least, no one was quite sure what had been discovered. People talked of a new particle "consistent with the Higgs boson," or said they had found "a Higgs boson," but not necessarily "the Higgs boson." Those phrases were helpful for rough-and-ready newspaper articles, but left questions over what scientists were sure of and what needed nailing down.

The bumps or "excesses" in the Atlas and CMS data pointed to a new particle that, once made, decayed immediately into other entities, including photons and electrons. These were the cleanest decay routes expected for the singular Higgs described by the Standard Model. So in that sense, the new particle behaves very much like the Standard Model Higgs boson.

Other details shed more light on the particle's identity. The mathematics of the Standard Model say that only a boson can decay into two clean photons. So the new particle is definitely a boson of some kind. Another feature is spin, a quantum property likened to the spin of a top. In the Standard Model, particles are either bosons or fermions. Force-carrying bosons, such as the photon and W and Z particles, are spin one. Fermions, such as quarks and electrons, are spin one-half. The Higgs particle is unique, so far at least, in being a spin-zero boson. The measurements of the new particle already rule out spin one (these particles cannot decay into two photons). All of which makes it likely that the new particle is indeed spin zero.

So what happens next? The new particle has to be characterized to within a whisker of its life, to see how it matches up to the Standard Model Higgs boson. Before the end of 2012, CERN physicists hope to confirm whether or not the new particle is spin zero. One way to do that is to look at how particles spray through the detectors. The decay products of a spin-zero Higgs boson are likely to leave tracks in the detectors that shoot equally in all directions, not some directions more than others. Other measurements will look at how often the particle is made in collisions, and in what proportions it decays through different channels. For example, a Standard Model Higgs boson weighing 125 GeV should decay into bottom quarks and antibottom quarks (their antimatter partners) around 57 percent of the time, and into two photons 0.2 percent of the time. Any discrepancies in these figures, which might take years to confirm, could open the door to a new realm of physics.

Many physicists dearly want to see something odd in the behavior of the new particle. As Steven Weinberg said (see Chapter 11), finding the Standard Model Higgs boson would not rescue physics from the doldrums, but put it into the doldrums. The reason is simple. Physicists have worked on the assumption that the Standard Model Higgs boson is real for decades. Finding it would give them no fresh clues on how to push beyond the Standard Model, to understand aspects of nature that it does not describe, such as gravity and dark matter. Any peculiar behavior of the new particle could signify something more exotic, perhaps showing the way to a more complete understanding of the world.

On July 4, it was too early to say whether the new particle displayed any strange characteristics. From the preliminary data it looked like the particle decayed into photons more often than expected, and less often into taus. The discrepancies could easily vanish with more data, but if not, they might be the lead physicists had long hoped for. One possibility, as described in Chapter 10, is that the new particle is one of a family of Higgs bosons described by the theory of supersymmetry. Asked about this, John Ellis said: "The mass is in a range where probably our vacuum would be unstable unless you add in some new physics, and that something else could be supersymmetry."

To give Atlas and CMS the best chance of answering at least some of these questions, CERN extended the LHC's proton-proton run by seven weeks in 2012, to mid-December. That may be enough to spot any glaring differences between the new particle and the Standard Model Higgs boson, but more subtle quirks in its behavior will have to wait until 2015 and beyond, when the LHC is scheduled to fire up at full design energy.

So it goes with science. No sooner is one question answered than another turns up to take its place. And another, and another. The process is surely endless. A clear and accurate description of reality lies somewhere over the horizon, and that may be out of reach forever. But the more scientists discover about the workings of nature, the closer we can come.

The rain is coming down so hard in Edinburgh that the words "James Clerk Maxwell Building" above the university's physics department are barely visible. Here, forty-eight years ago, Peter Higgs first described a new particle after it dropped out of equations he wrote down with pencil and paper. Two weeks after the CERN announcement, he is here again, in dark blue trousers and a corduroy shirt, and looks good for his eighty-three years. The only obvious frustration of age seems to be twin hearing aids, one of which is mistuned and shrill, the other silenced by a dead battery. "Speak into this ear," he says, pointing.

For Higgs, the July 4 meeting was emotional, but what lay behind his reaction was never quite clear on that day. Sitting at a desk with a laptop, he is now visibly moved to watch the closing minutes of that morning's seminars again. The video pauses at an unfair moment, when Higgs has shed a tear or two and looks almost distraught, a curious manifestation of human joy, made all the more conspicuous by the sea of gleeful faces around him. How did he feel at that precise moment? "I was about to burst into tears. I was knocked over by the wave of the reaction of the audience. Up until then I was holding back emotionally, but when the audience reacted I couldn't hold back any more. That's the only way I can explain it," he said.

The idea Higgs scribbled down nearly half a century ago became much more than a few penciled equations in a notebook. The theory itself gained weight as physicists in the twentieth century found it had real purpose, verified it mathematically, and uncovered indirect evidence that it described nature accurately. At first an obscure note in a journal, the theory acquired an importance that justified a decades-long search, with multibillion-dollar machines, that became the careers, dreams, and lives of thousands of scientists and engineers. This was their success. The tears Higgs cried were not for what he contributed, but for what that contribution had become to others.

The intensity of the scientific process can be all consuming. Follow the work long enough, and it can seem an end in itself: to make exquisitely complex machines, to operate them well, and to pull from its noisy data some obscure but meaningful signal. All these things are extraordinarily valuable. They develop skills and push technology. But to focus on them is to lose sight of the original goal of the enterprise. These experiments dig out the workings of nature at a fundamental level. What do we know now that we didn't know before? Forget the particle for a moment. That was only ever a means to an end, the smoking gun that confirmed something far more fascinating. We know now that the universe is filled with a field that makes massless particles massive. We understand a little more about how this all came to be.

In the physics department cafeteria, Higgs picked up a plate of chicken stir-fry and a can of fizzy apple drink and queued for the cashiers behind postdocs and lecturers. He then joined a table with Alan Walker and some of the Edinburgh Atlas group to discuss what the new particle might mean for physics and the years of work that lie ahead. Since the discovery at CERN, a flood of emails and letters had arrived for him, which covered the living-room floor of his apartment. Some had to be answered immediately, but others would have to wait. After lunch, Higgs headed down to the foyer, collected his things, and said a handful of goodbyes. At least the attention would die down soon, he said. And with that, he pulled on his coat, looked up at the sky, and walked out into the rain.

ACKNOWLEDGMENTS

This book came about because physicists took the time to share their passion with me. It was their enthusiasm that led to my fascination with particle physics and to my interest in the story of the Higgs particle. Thousands of people have played a part in the story over the years; told in full, it would fill a library. I have been able to speak with only a minority of those involved, and this is just one version of events.

I would like to thank all of those who got behind the idea for this book and helped to bring it to fruition. My agent, Peter Tallack at The Science Factory, was a great supporter of the idea from the start, and his advice and cajoling were invaluable. I am indebted to my editors at Virgin, Ed Faulkner and Davina Russell, who were excited enough about the CERN story to publish the book. Thanks also to Lara Heimert, my U.S. editor at Basic Books, whose insightful comments and guidance helped enormously. Her team at Basic Books, including the copyeditor, Kathy Streckfus, improved the book immeasurably.

At the *Guardian*, Nick Hopkins granted me a leave of absence, which I greatly appreciate. The leave allowed me to escape the crush of deadlines and distractions long enough to complete the manuscript. I would also like to say a special thanks to my colleagues Alok Jha, David Adam, Karen McVeigh, and James Kingsland for covering for me.

Countless scientists and engineers gave up some of their precious time to talk with me while I was researching the book, and I'm profoundly grateful to all of them. The end result was vastly improved thanks to those who checked my clumsy drafts, including Steven Weinberg, John Ellis, Michael Fisher, Lyn Evans, John Conway, Gerry Guralnik, and Dick Hagen. Peter Higgs provided comprehensive and invaluable comments on key chapters and put me right on many occasions. His help is a debt I cannot repay.

Thanks to Freeman Dyson for digging back through his memories to tell me about Peter Higgs's visit to the Institute for Advanced Study in 1966 and for his reflections on Robert Oppenheimer. For explaining their contributions to the theory of the origin of mass, thanks to the six men who came up with the idea: François Englert, Robert Brout, Peter Higgs, Gerry Guralnik, Dick Hagen, and Tom Kibble.

The archivists at King's College London made tracking down old minutes of the Maxwell Society more fun than I could have imagined possible. Thanks also to the curators of the Ava Helen and Linus Pauling Papers at the Oregon State University Libraries Special Collection for their mind-boggling efficiency in providing letters between Higgs and Linus Pauling. Those in charge of the Bertrand Russell archive at McMaster University were similarly impressive. The staff at the science reading room of the British Library were unfailingly competent and kind in responding to my requests for help in searching for obscure research papers and U.S. congressional hearings.

Ben Allanach at Cambridge University and Gian Francesco Giudice at CERN did better than most in helping me understand the Higgs mechanism. I am very grateful to both of them. Thanks are also due to Alan Guth and Paul Steinhardt, who explained crucial aspects of inflation to me.

Steven Weinberg was kind enough to take me through his Nobel Prize–winning work and made sure I didn't miss my plane. Gerardus 't Hooft helped me understand his work on the Higgs mechanism, and Martinus Veltman explained his own theory of why the Higgs particle might turn out to be invisible, or to simply not exist.

I'm grateful to Frank Wilczek and John Marburger III for taking my questions about doomsday scenarios seriously, and also to Stephen

Hawking and Gordon Kane for sharing their thoughts on whether the Higgs particle will ever be found. The high-energy physics group at University College London, including David Miller, Jonathan Butterworth, Mark Lancaster, and Nikos Konstantinidis, all provided great tales and sensible guidance, and often both at once.

Alvin Trivelpiece could not have helped more in piecing together the story of the Superconducting Supercollider. His deep knowledge and insight were crucial.

At Fermilab, Kurt Riesselmann helped to arrange interviews, while the deeply knowledgeable Adrienne Kolb gave me fantastic access to the laboratory's extensive archives. The Fermilab scientists Chris Quigg, Leon Lederman, Dmitri Denisov, Robert Roser, Jacobo Konigsberg, and John Conway were all generous with their time and advice.

At CERN, James Gillies and others in the communications office also helped to arrange interviews and, on occasion, ensured that I had a roof over my head. Existing or former staff scientists at CERN were unfailingly helpful and patient. I am particularly grateful to Jos Engelen, Luciano Maiani, Herwig Schopper, Chris Tully, Roberto Tenchini, Fabiola Gianotti, Jim Virdee, John Swain, John Ellis, and Patrick Janot. Special thanks to Lyn Evans and Roger Cashmore for taking so much time to explain details of science and engineering.

When I began work on the book I had not heard of the Higgs portal or hidden worlds. My gratitude goes to James Wells at CERN and John March-Russell at Oxford University for explaining the idea and giving me something profound to be excited about.

To my friends and family whom I've neglected and bored in equal measure: thanks for still being around. Finally, my sincere thanks to Jo Marchant, whose patience and advice kept me afloat; and to Poppy, who makes finishing all the more worthwhile.

NOTES

Some of the articles mentioned here are freely available on a preprint server called "arXiv," an online repository of academic papers in mathematics and physics hosted by Cornell University Library. To find, for example, Stephen Hawking's 1995 paper "Virtual Black Holes," type the article ID, hep-th/9510029v1, into the search box on the arXiv website (www.arxiv.org) and it should take you straight to the article. Many of the articles later appeared in journals; in such cases, I have supplied the journal citations as well as the arXiv article ID numbers.

Chapter 1

1. Higgs was invited to the University of North Carolina at Chapel Hill by Bryce de Witt, who became a champion of the many-worlds interpretation of quantum mechanics developed by Hugh Everett III in the 1950s. The university was founded in the 1780s, making it the oldest state university in the United States.

2. For a description of the theologian's thinking on mass and volume, see *Medieval Cosmology: Theories of Infinity, Place, Time, Void, and the Plurality of Worlds*, by Pierre Duhem, University of Chicago Press, 1987.

3. See *Cosmology: The Science of the Universe*, by Edward Robert Harrison, Cambridge University Press, 2000.

4. Published in Latin under the title *Philosophiæ Naturalis Principia Mathematica*, and usually called simply the *Principia* or the *Principia Mathematica*, Newton's treatise consists of three books. The title may be translated as "Mathematical Principles of Natural Philosophy."

5. Written in English, *Opticks* was Newton's second great book and covers theories of reflection, refraction, and color.

6. For a full history, see *J. J. Thomson and the Discovery of the Electron*, by E. A. Davis, Taylor & Francis, 1997. The date of discovery is contested by Abraham Pais in *Inward Bound*, Clarendon Press, 1988.

7. For an account of Rutherford's demolition of Thomson's atomic model, see *The Last Sorcerers: The Path from Alchemy to the Periodic Table*, by Richard Morris, Joseph Henry Press, 2003.

8. No atom is simpler than hydrogen. It contains a nucleus of one proton, which is circled by one electron. The next most simple atom, helium, has two protons, two neutrons, and two electrons.

9. For further notes, see *The Neutron and the Bomb: A Biography of Sir James Chadwick*, by Andrew Brown, Oxford University Press, 1997.

10. For more on the history of quarks, see *The Quantum Quark*, by Andrew Watson, Cambridge University Press, 2004, and *Quarks and Gluons*, by M. Y. Han, World Scientific, 1999.

11. For a comprehensive guide to these, try *The Particle Odyssey: A Journey to the Heart of Matter*, by Frank Close, Michael Marten, and Christine Sutton, Oxford University Press, 2004.

12. A comprehensive history of the discoveries that underpin the Standard Model can be found in Lillian Hoddeson et al., eds., *The Rise of the Standard Model: Particle Physics in the 1960s and 1970s*, Cambridge University Press, 1997, 199.

13. Matter particles are divided into three generations that differ only by their mass. The first-generation quarks are up and down, the second generation includes charm and strange, and the third generation includes top and bottom. In a sense, the second- and third-generation quarks are more massive cousins of the first-generation quarks. The first generation of leptons includes the familiar electron and the electron neutrino. The second, heavier generation includes the muon and the muon neutrino. The third generation of leptons comprises the tau and the tau neutrino. The muon and tau are heavier versions of the electron.

14. The force-carrying particles of the Standard Model are bosons and include the photon (electromagnetic force), the gluon (strong force), and the W and Z (the weak force). There is a fifth boson, which is the Higgs particle. Bosons are named after the Indian physicist Satyendra Nath Bose and Albert Einstein. For more on Bose, see *Satyendra Nath Bose: His Life and Times*, edited by Kameshwar C. Wali, World Scientific, 2009.

15. Of the fundamental forces of nature, the weak force is probably the least well known. All particles except gluons and photons feel the weak force. It acts over such short distances that it effectively acts on contact. The weak force is involved in radioactive beta decay, in which radioactive elements emit high-energy electrons or positrons. Quarks can change from one variety to another, in a process known as "flavor changing," by exchanging a W boson.

16. Newton's laws of motion work fine for describing objects (or particles) that are moving much slower than the speed of light. But close to the speed of

light, the laws change dramatically and Einstein's special theory of relativity takes over. The theory is a consequence of two statements: first, that the speed of light is the same for all onlookers, regardless of their relative velocities; and second, that the laws of physics are the same in all inertial (not accelerating) frames of reference. Put another way, the laws of physics look the same whether your lab is stationary or hurtling through space at a constant velocity.

17. Scientists estimate that the universe cooled enough for the Higgs field to switch on just 1 picosecond, or a trillionth of a second, after the Big Bang.

18. Scientists generally agree that the universe is 13.7 billion years old. What happened before that? Theories so far have nothing to say on the matter, and we might never know. Stephen Hawking has compared the question of what happened before the Big Bang with asking: What's north of the North Pole?

19. Removing or altering the strength of the Higgs field would have dramatic implications. One example is found in chemistry. The Higgs field is a way for the electron to acquire mass. Without the Higgs field, electrons would remain massless, and they would move too fast to be captured in orbit around atomic nuclei. That would be curtains for the periodic table of elements as we know it.

20. The Standard Model Higgs field is what is known as a complex field. It consists of two neutral and two charged component fields. The two charged components give mass to the positively and negatively charged W bosons. One neutral component gives mass to the Z boson. The Higgs boson is the quantum of the remaining neutral component field.

21. For more on Gödel's work, see *Thinking about Gödel and Turing: Essays on Complexity, 1970–2007,* by Gregory J. Chaitin, World Scientific, 2007.

22. For more on von Neumann's work on game theory, see *Prisoner's Dilemma: John von Neumann, Game Theory and the Puzzle of the Bomb,* by William Poundstone, Anchor Books, 1993.

23. Interview with the author, August 2008.

24. Interview with the author, August 2007.

25. As recalled in an interview with the author, August 2008.

26. See "Conserved Currents and Associated Symmetries: Goldstone's Theorem," by Daniel Kastler, Derek W. Robinson, and André Swieca, *Communications in Mathematical Physics* 2, no. 2 (1966): 108–120.

27. Oppenheimer stood down from his directorship in 1966 after being diagnosed with throat cancer. He died the following year, on February 18, 1967, at the age of sixty-two.

28. Interview with the author, August 2008.

29. See "SBGT and All That," by Peter Higgs, in *Weak Neutral Currents,* edited by David B. Cline, Westview Press, 1997.

Chapter 2

1. For a full account of Melba's Chelmsford broadcast, see *A Concise History of British Radio 1922–2002: 80 Years of Key Developments*, by Sean Street, Kelly Publications, 2002, and *The Emergence of Broadcasting in Britain*, by Brian Hennessy, Southerleigh, 2005.

2. For more on Maxwell's unconventional appearance, see, for example, *The Atomic Scientists: A Biographical History*, by Henry Abraham Boorse, Lloyd Motz, and Jefferson Hane Weaver, Wiley, 1989, and *The Demon in the Aether: The Story of James Clerk Maxwell*, by Martin Goldman, Adam Hilger, 1984.

3. See, for example, *The Gateway to Understanding: Electrons to Waves and Beyond*, by Matthew Radmanesh, AuthorHouse, 2005, and *Scientific American Inventions and Discoveries: All the Milestones in Ingenuity from the Discovery of Fire to the Invention of the Microwave Oven*, by Rodney Carlisle, Wiley, 2004. In 1845 Faraday became the first to use the term "magnetic field."

4. See, for example, *Einstein's Essays in Science*, by Albert Einstein, Dover Publications, 2009.

5. The concept of unification has become a powerful guide in physics and other sciences. For more on the great unifications in physics, see *Ideals and Realities: Selected Essays of Abdus Salam*, by Abdus Salam, edited by Z. Hassan and C. H. Lai, World Scientific, 1984, and *Strange Matters: Undiscovered Ideas at the Frontiers of Space and Time*, by Tom Siegfried, Joseph Henry Press, 2002.

6. For more on the ether, see *Quintessence: The Mystery of Missing Mass in the Universe*, by Lawrence Krauss, Vintage, 2001, and *From Newton to Hawking: A History of Cambridge University's Lucasian Professors of Mathematics*, by Kevin C. Knox and Richard Noakes, Cambridge University Press, 2002.

7. See, for example, *The Cambridge Companion to Newton*, by I. Bernard Cohen and George E. Smith, Cambridge University Press, 2002.

8. For more on Kelvin's predictions, see, for example, *The New Physics for the Twenty-First Century*, by Gordon Fraser, Cambridge University Press, 2006, and *Springs of Scientific Creativity: Essays on Founders of Modern Science*, by Ted Davis, Roger H. Stuerwer, and Rutherford Aris, University of Minnesota Press, 1983.

9. See *Men Who Made a New Physics: Physicists and the Quantum Theory*, by Barbara Lovett Cline, University of Chicago Press, 1987.

10. Quoted in *Great Physicists: The Life and Times of Leading Physicists from Galileo to Hawking*, by William H. Cropper, Oxford University Press, 2001, which references *Scientific Autobiography and Other Papers* (Physikalische Abhandlungen und Vorträge), by Max Planck, New York Philosophical Library, 1949.

11. For more on Einstein's quantum description of the effect, see *Einstein, Bohr and the Quantum Dilemma: From Quantum Theory to Quantum Informa-*

tion, by Andrew Whitaker, Cambridge University Press, 2006, and *Subtle Is the Lord: The Science and the Life of Albert Einstein,* by Abraham Pais, Oxford University Press, 2005.

12. Gases absorb and emit different colors of light because of the differing energies of their electron orbits. You can think of the orbits as concentric rings around the atomic nucleus. An electron will be pushed up into a higher orbit if it absorbs enough energy. The wavelength of light absorbed is governed by the energy gap between the orbits. When the electron falls back down again, it releases the same wavelength of light. By studying the "team colors" of gases, scientists can learn how the arrangements of their electron orbits differ.

13. For Heisenberg's thoughts on his time on Heligoland, see *Truth and Beauty: Aesthetics and Motivations in Science,* by Subrahmanyan Chandrasekhar, University of Chicago Press, 1990.

14. See, for example, *Facts and Mysteries in Elementary Particle Physics,* by Martinus Veltman, World Scientific, 2003.

15. For more on Schrödinger's drawing on de Broglie's idea and formulation of quantum theory, see *It Must Be Beautiful: Great Equations of Modern Science,* by Graham Farmelo, Granta, 2003.

16. Dirac read Schrödinger's first paper on wave mechanics and didn't like it. He preferred Heisenberg's matrix mechanics and drew on it to set up a general formulism of quantum mechanics in his Ph.D. thesis.

17. Several books describe the nature of antimatter and its initial discovery. The recent book *Antimatter,* by Frank Close, Oxford University Press, 2009, is hard to beat as an introduction.

18. According to Higgs, Dirac "got by on his mathematics, but was hopeless at practical engineering." Peter Higgs, letter to author, March 2010.

19. See "Verifying the Theory of Relativity," by S. Chandrasekhar, in *Bulletin of the Atomic Scientists,* June 1975. By the time Higgs became a student at King's College London, he had decided Eddington was "a crackpot." Eddington's attempt to unify quantum theory with relativity and gravitation, known as his "fundamental theory," was described by Higgs as "total crap."

20. With the help of Danish fishermen, Niels Bohr was spirited out of Denmark aboard an RAF mosquito bomber and taken via England to Los Alamos National Laboratory to join the Manhattan Project, the U.S. atomic bomb effort. For a full account and far more on Bohr, see *Thirty Years That Shocked Physics,* by George Gamow, Dover Publications, 1985.

21. Much of the relevant work on fission was done by Hahn with his colleagues Lise Meitner and Fritz Strassman. For more detail, see *The Harvest of a Century: Discoveries in Modern Physics in 100 Episodes,* by Siegmund Brandt, Oxford University Press, 2009.

22. The extraordinary and often moving dialogue between the interned physicists is compiled in *Operation Epsilon: The Farm Hall Transcripts*, by Charles Frank, University of California Press, 1993.

23. Freeman Dyson, quoted from *The Day After Trinity*, a documentary about the life and work of Robert Oppenheimer. Directed and produced by Jon Else and KTEH public television, San Jose, California, and broadcast on PBS in April 1981.

Chapter 3

1. This was Higgs's presidential address to the society. The minutes of the meeting are held in an archive at King's College London.

2. For more on solipsism and Descartes's philosophical musings, see, for example, *Human Knowledge: Its Scope and Limits*, by Bertrand Russell, Routledge, 2009.

3. The late science fiction author was a King's graduate and studied at the university a few years before Higgs enrolled. His ideas on interplanetary travel were widely derided at the time.

4. Several excellent books have been written about Feynman. For his own recollection of the events leading to renormalization theory, his Nobel lecture (delivered on December 11, 1965) is hard to beat.

5. This is not intended as a slight on Feynman's character. Accounts of Feynman's relationships and marriages are rich with tenderness and devotion. For more about Feynman, see *The Great Physicists: The Life and Times of Leading Physicists from Galileo to Hawking*, by William H. Cropper, Oxford University Press, 2001.

6. In the 1950s, postgraduate fellowships were a rarity and extremely prestigious. In London, only a few students were nominated for fellowships each year. Higgs graduated from King's College as the top physicist in his year.

7. Nambu published several preliminary articles in 1960 in, for example, *Physical Review Letters and Proceedings of the 10th Annual Rochester Conference on High Energy Nuclear Physics*. A more complete account can be found in *Physical Review* 122, no. 1 (1961): 345–358.

8. Bruno Zumino at the University of California at Berkeley once recounted his attempts to understand Yoichiro Nambu. He said: "I had the idea that if I can find out what Nambu is thinking about now, I'll be ten years ahead in the game. So I talked to him for a long time. But by the time I figured out what he said, ten years had passed." Similarly, Ed Witten at the Institute for Advanced Study in Princeton has said of Nambu: "People don't understand him, because he is so far-sighted."

9. Superconductivity is a far more complex and subtle process than I have described here. For a good introduction to the subject, see *The Rise of the Su-*

perconductors, by P. J. Ford and G. A. Saunders, CRC Press, 2004, or *The Early Universe*, by Gerhard Börner, Springer, 2004.

10. There is a helpful section on broken symmetries in *New Theories of Everything*, by John Barrow, Oxford University Press, 2007. See also *The Second Creation: Makers of the Revolution in Twentieth-Century Physics*, by Robert Crease and Charles Mann, Rutgers University Press, 1996, and *Broken Symmetries, A Scientific Backgrounder on the Nobel Prize in Physics*, Royal Swedish Academy of Sciences, 2008.

11. These particles, named "Goldstone bosons," were originally thought to plague any system where the symmetry was spontaneously broken. The problem was this: massless particles would take very little energy to make, so if they existed, they should be as easy to observe in nature as the photons that make up light. The fact that no one had seen these massless particles suggested very strongly that there was no spontaneous symmetry-breaking mechanism that gave rise to particle masses. The problem became known as the "Goldstone theorem."

12. Gilbert was Guralnik's Ph.D. supervisor.

13. Prentki was head of the theory division at CERN.

14. Some years later, Yoichiro Nambu revealed to Higgs that he was the referee in question.

15. For full source information on the key papers from each of the three teams, see the Bibliography.

16. The Higgs field generates mass in different ways, depending on the type of particle. In the Standard Model, there are four varieties of force-carrying particles: the photon, the gluon, the W boson (which can be positive or negative), and the Z boson. Before the Higgs field activates, all of these particles are massless, and the waves associated with them oscillate only in the transverse plane (the plane perpendicular to the direction a particle is traveling in). Photons and gluons are not affected by the Higgs field, but when the W and Z particles interact with the Higgs field, their waves gain a longitudinal component (that is, they can oscillate in the direction of travel). It is this extra degree of freedom that makes the W and Z particles massive. For quarks and leptons (except perhaps neutrinos), the situation is different. Before the Higgs field goes to work, quarks and electrons have what is called a "single spin state." Some spin in the direction they move in, while others spin in the opposite direction. When the Higgs field switches on, these particles get both spin states, and it is this process that is thought to give them mass.

Chapter 4

1. In *Ideals and Realities: Selected Essays of Abdus Salam*, by Abdus Salam, edited by Z. Hassan and C. H. Lai, World Scientific, 1984, Salam describes how

science thrives on the interchange of ideas and continuous criticism. Some physicists, such as Lee Smolin at the Perimeter Institute in Ontario and Peter Woit at Columbia University, have made themselves unpopular by suggesting that a major area of physics is failing today by focusing too much on string theory, a description of nature that envisages particles as tiny vibrating strings of energy.

2. For more information, see, for example, *Fundamental Forces of Nature: The Story of Gauge Fields*, by Kerson Huang, World Scientific, 2007, in which the author, an emeritus professor of theoretical physics at the Massachusetts Institute of Technology, states: "There was a revolt against quantum field theory, perhaps out of disillusionment." See also *50 Years of Yang-Mills Theory*, edited by Gerardus 't Hooft, World Scientific, 2005, and the chapter on Julian Schwinger and relativistic quantum field theory in *Nobel Lectures in Physics, 1963–1970*, edited by Stig Lundqvist, World Scientific, 1998.

3. For more on the development of S-matrix, see *Pions to Quarks: Particle Physics in the 1950s*, by Laurie Mark Brown, Max Dresden, and Lillian Hoddeson, Cambridge University Press, 1989. See also *Hyperspace: A Scientific Odyssey Through Parallel Universes, Time Warps, and the Tenth Dimension*, by Michio Kaku, Oxford University Press, 1994.

4. "Black box" is the phrase used by Higgs to describe S-matrix in an interview with the author in 2008. Higgs is referring to the scientific meaning, which is a system or device where all that is known are the inputs and outputs, and not the workings that relate the two. This is not to be confused with the kind of black box that keeps aircraft accident investigators in business.

5. Particles are described by a number of characteristics, but among the most common are mass and charge. Theories that predict particle masses are particularly helpful because physicists know how much energy will be required to make one in a particle accelerator, since a greater mass corresponds to a higher energy. The mass of an unstable particle helps scientists work out what stable particles it will decay into once created. These decays are often used as evidence of a particle having been made.

6. Think of the Earth's magnetic field. At every point on the Earth's surface (and in the atmosphere), it has a direction and a strength, with the field being stronger at the poles. The same goes for gravity. Fields that have a strength and direction are known as "vector fields." Now think of temperature. The temperature varies enormously around the globe, but a "temperature field" has no direction. The same is true for the Higgs field. These kinds of fields, which have strength but no direction, are called "scalar fields."

7. Were it possible to alter the strength of the Higgs field, it would have a direct impact on the masses of electrons inside atoms, which in turn would change the sizes and stability of atoms.

8. The article is republished in a collection of essays, *Facing Up: Science and Its Cultural Adversaries*, by Steven Weinberg, Harvard University Press, 2001.

9. S. Weinberg, "A Model of Leptons," *Physical Review Letters* 19, no. 21 (1967): 1264–1266.

10. This is an understatement: it was Glashow's Nobel lecture, given on December 8, 1979. He won the prize for his work on unifying the electromagnetic and weak forces.

11. See G. S. Guralnik, "The History of the Guralnik, Hagen and Kibble Development of the Theory of Spontaneous Symmetry Breaking and Gauge Particles," *International Journal of Modern Physics A* 24, no. 14 (2009): 2601–2627.

12. For this rich picture of Utrecht and the dual use of its physics department, see Gerardus 't Hooft's autobiography in *Les Prix Nobel* (The Nobel Prizes), 1999, published by the Nobel Foundation.

13. For a brief autobiography of Veltman, see *Les Prix Nobel*, 1999.

14. As described by Gerardus 't Hooft in an interview with the author in 2009.

15. As quoted in *The Nobel Prize: A History of Genius, Controversy and Prestige*, by Burton Feldman, Arcade Publishing, 2001.

16. In 2010, all six physicists were awarded the J. J. Sakurai Prize for Theoretical Particle Physics in recognition of their work on the origin of mass. The award ceremony, which took place in February 2010 in Washington, D.C., was expected to be the first time all six would meet, but Higgs was unable to attend.

17. See Guralnik, "The History of the Guralnik, Hagen, and Kibble Development."

18. See "SBGT and All That," by Peter Higgs, published in *Weak Neutral Currents*, edited by David B. Cline, Westview Press, 1997.

Chapter 5

1. Two detailed accounts of the discovery of neutral currents were particularly helpful. The first was written by Donald Perkins, who was at the heart of the action. See Chapter 25, "Gargamelle and the Discovery of Neutral Currents," in *The Rise of the Standard Model*, edited by Lillian Hoddeson et al., Cambridge University Press, 1997. The second useful account, "The Discovery of Neutral Currents," by Peter Galison, appears in *Weak Neutral Currents*, edited by David B. Cline, Westview Press, 1997.

2. A great account of the physicists' work appears in *The Particle Odyssey: Journey to the Heart of Matter*, by Frank Close, Michael Marten, and Christine Sutton, Oxford University Press, 2004.

3. For details, see *Rutherford: Recollections of the Cambridge Days*, by Sir Mark Oliphant, Elsevier Science, 1972.

4. See *Lawrence and His Laboratory: A History of the Lawrence Berkeley Laboratory*, vol. 1, by J. L. Heilbron and Robert W. Seidel, University of California Press, 1989.

5. See, for example, *The Quark Machines: How Europe Fought the Particle Physics War*, by Gordon Fraser, Taylor & Francis, 1997.

6. Abraham Pais describes the rise of U.S. physics immediately after World War II in *Inward Bound: Of Matter and Forces in the Physical World*, Oxford University Press, 1986. He notes, on p. 473, that "European science and technology could keep in step with developments elsewhere and that the brain drain could be slowed down only by joining forces."

7. The CERN photo archive has a wonderful photograph of Adams wielding the spent item. See the CERN Document Server, Record 39074.

8. For a comprehensive portrait of Robert Wilson and his work at Fermilab, see *Fermilab: Physics, the Frontier and Megascience*, by Lillian Hoddeson, Adrienne W. Kolb, and Catherine Westfall, University of Chicago Press, 2008.

9. For an enjoyable account of this exchange, see *The Second Creation: Makers of the Revolution in Twentieth-Century Physics*, by Robert P. Crease and Charles C. Mann, Macmillan, 1986.

10. For this and more of Wilson's quirks, see Hoddeson et al., *Fermilab*.

11. Neutrinos are not affected by the strong force or the electromagnetic force, and, since they have such a tiny mass, they are barely influenced by gravity. They do feel the weak force, though, which makes them ideal particles for the study of electroweak theory.

12. See "The Discovery of Neutral Currents," by Peter Galison, in *Weak Neutral Currents*, edited by David B. Cline, Westview Press, 1997.

13. For original accounts of this episode, see Galison, "The Discovery of Neutral Currents," and "Distrust and Discovery: The Case of the Heavy Bosons at CERN," by John Krige, *Isis* 92, no. 3 (2001): 517–540.

14. Hoddeson et al., *Fermilab*, 166–167.

15. See *Selected Papers of Abdus Salam*, by Abdus Salam and Ahmed Ali, World Scientific, 1994.

16. See Galison, "The Discovery of Neutral Currents."

17. "The Feynman Lectures on Physics: Exercises," Addison-Wesley, 1965.

18. See Krige, "Distrust and Discovery."

19. The CERN archive holds this memo from John Adams, entitled "Approval of ppbar Facility," to C. Rubbia and S. van der Meer, dated June 8, 1978.

20. The CERN archive holds this corrigendum from Adams, entitled "Talk to CERN Staff," dated August 24, 1978.

21. See Hoddeson et al., *Fermilab*.

22. Letter from the CERN archive, dated December 20, 1982, copied to Sir Alec Morrison, president of the CERN council.

23. See Krige, "Distrust and Discovery."

24. Several other articles that cover aspects of the search for W and Z particles include "The W and Z at LEP," by Christine Sutton at CERN and Peter Zerwas at the German laboratory DESY, in the *CERN Courier*, May 2004, and "The W and Z Particles: A Personal Recollection," by Pierre Darriulat, in the *CERN Courier*, April 2004.

25. Krige, "Distrust and Discovery." In Krige's account, he states that di Lella's CERN colleague Pierre Darriulat verified this version of events. In an interview, di Lella told Krige: "I know with Carlo, Carlo will beat you and he will use all kinds of tricks to beat you. You have to accept the man for what he is." Rubbia declined to give an interview for this book.

26. See "A Phenomenological Profile of the Higgs Boson," by J. Ellis, M. K. Gaillard, and D. V. Nanopoulos, *Nuclear Physics B* 106, no. 2 (1976): 292–340.

27. See "Reagan Aide Sees Physics Lag," *New York Times*, April 20, 1983.

28. *New York Times* editorial, June 6, 1983.

Chapter 6

1. Much of this account comes from two lengthy interviews I had with Trivelpiece, who also provided me with copies of several in-depth lectures he had given on the Supercollider story.

2. For a detailed explanation of the Supercollider's capabilities and the centrality of the question of the origin of mass, see "The Superconducting Supercollider," by J. David Jackson, Maury Tigner, and Stanley Wojcicki, *Scientific American*, March 1986.

3. See "White House, DOE Announce Support for SSC," in *Ferminews*, the Fermilab in-house journal, February 13, 1987.

4. See *Fermilab: Physics, the Frontier and Megascience*, by Lillian Hoddeson, Adrienne W. Kolb, and Catherine Westfall, University of Chicago Press, 2008.

5. Lederman's thoughts on Rubbia's selling of the LHC appear in "Collision over the Supercollider," by Gary Taubes, *Discover*, July 1985.

6. For the comments attributed to James P. Sethna, see Krumhansl's obituary ("James Krumhansl, 84, Opponent of Supercollider"), *New York Times*, May 22, 2004.

7. Krumhansl's prediction is covered in the *Physics Today* "Washington Report," August 1987, 50.

8. Secretary Herrington elaborated on this point in a hearing before the House of Representatives Subcommittee on Energy Research and Development of the Committee on Science, Space and Technology held on March 17

and 18, 1987. In an exchange with the Republican Sherwood Boehlert, he said: "It is an American project involving American high-energy physics, and it is intended to maintain American leadership in this particular field."

9. The late governor of Texas was allegedly given a choice between supporting the supercollider or the space station. She chose not to back the Supercollider. From notes on a lecture given by Alvin Trivelpiece at the American Physical Society meeting in Tampa, Florida, April 19, 2005.

10. Lederman's publisher argued that no one had ever heard of Peter Higgs, so using his name in the book title was out. Every physicist I talked to about the name "the God particle" hates it. The main objections are that it seems grandiose and potentially offensive to the religious. But there are other complaints. Rightfully, physicists say the term is meaningless and ridiculous. Other particle names may sound obscure, but in many cases they make sense. The naming was particularly unfortunate because Lederman's book appeared at a time when creationism was becoming a greater concern for many U.S. schools. For a glimpse of the issues journalists face when mentioning the nickname, see "What's in a Name? Parsing the God Particle, the Ultimate Metaphor," by Dennis Overbye, *New York Times*, August 7, 2007.

Chapter 7

1. For a comprehensive guide to the design and construction of the Large Electron-Positron collider, see *LEP: The Lord of the Collider Rings at CERN, 1980–2000*, by Herwig Schopper, Springer, 2009. CERN's in-house journal, the *CERN Courier*, is a useful and free-to-access resource that is particularly helpful on the machine.

2. The energy that electrons lose when they follow a curved path is called "synchrotron radiation." This lost energy is an annoyance for making high-energy beams, because the faster the electrons are traveling, the more radiation they emit, so more energy has to be put in from the accelerating equipment. There is another side to the coin, though. Synchrotron X-rays are incredibly intense and so are useful for studying all manner of objects, including proteins and aircraft engines. At the Diamond synchrotron facility in Oxfordshire, synchrotron radiation has been used to examine fragments of the Dead Sea scrolls without unraveling the delicate material.

3. The principle of owning the land beneath one's home to the center of the Earth can be traced back to William Blackstone, an eighteenth-century English judge, who declared the maxim in *Commentaries on the Laws of England*. He used the Latin phrase, "Cuius est solum, eius est usque ad caelum et ad inferos," meaning "For whoever owns the soil, it is theirs up to Heaven and down to Hell."

4. From Margaret Thatcher's speech to the Royal Society, Fishmongers' Hall, City of London, September 27, 1988.

5. The effect of the Earth tides on LEP was covered by the *New York Times* in "Moon Is Blamed for Blips in Particle Accelerator," by Malcolm W. Browne, November 27, 1992. A more detailed paper was written up at CERN: "Effects of Terrestrial Tides on the LEP Beam Energy," by L. Arnaudon et al., March 2, 1994.

6. Jack Steinberger, quoted in the CERN document "The Aleph Experience: 25 Years of Memories," 2nd ed., January 2006.

7. Maddox was a brilliant editor and I do not wish this to sound too harsh. The editorial in question appeared in *Nature* 362 (April 29, 1993): 785.

8. Some scientists think the true Higgs field may have powered *inflation*, the exponential expansion of the early universe. See, for example, "The Standard Model Higgs Boson as the Inflaton," by F. Bezrukov and M. Shaposhnikov, *Physics Letters B* 659, no. 3, (January 24, 2008): 703–706.

9. The story was widely reported at the time. In the United Kingdom, the *Independent* declared in its headline: "Marital Row Blows Fuse on Big Bang Theory."

10. For more on the beer bottle incident, see CERN, "The Aleph Experience," and "Two Green Bottles Leave Physicists Hanging," by Declan Butler, *Nature* 381 (June 27, 1996).

Chapter 8

1. This profoundly implausible scenario is one of the more outlandish ideas that physicists had to review when a small group of protesters raised concerns over the safety of particle colliders at the end of the twentieth century.

2. See "A Little Big Bang," by M. Mukerjee, *Scientific American*, March 1999.

3. Sadly, though understandably, neither *Scientific American* nor Wilczek were able to find the original text of the letter.

4. In January 1997, a radioactive form of hydrogen, tritium, was detected in groundwater south of the High-Flux Beam Reactor at Brookhaven National Laboratory. The tritium concentrations exceeded state and federal drinking-water standards, but they were confined to the Brookhaven site. The Department of Energy decided to close the reactor permanently in 1997, two years before controversy broke over the Rick collider.

5. See "The Case of the Deadly Strangelets," by Robert Crease, *Physics World*, July 2000, 19–20. Crease quotes a physicist's response to the question: "That connection would not, ah, have occurred to me."

6. For a full account of the story, see *Polywater*, by Felix Franks, MIT Press, 1981.

7. See "Accelerator Disaster Scenarios, the Unabomber, and Scientific Risks," by Joseph Kapusta, in *Physics in Perspective*, Springer, 2008.

8. Several papers and reports were helpful on this issue. Most relevant are: "Review of the Safety of LHC Collisions," by the CERN LHC safety assessment group, J. Ellis, G. Giudice, M. Mangano, I. Tkachev, and U. Wiedemann, *Journal of Physics G* 35, no. 11 (2008): 115004–115021, and a later report entitled "Study of Potentially Dangerous Events During Heavy Ion Collisions at the LHC: Report on the LHC Safety Study Group," CERN document, February 28, 2003. A review released by Brookhaven gives a detailed analysis of the issues. "Review of Speculative 'Disaster Scenarios' at RHIC" appeared in original form on September 28, 1999, and in revised form on July 14, 2000.

9. See "Gravitational Effects on and of Vacuum Decay," by Sidney Coleman and Frank de Luccia, *Physical Review D* 21, no. 12 (1980): 3305–3315.

10. See "Is Our Vacuum Metastable," by Michael S. Turner and Frank Wilczek, *Nature* 298 (August 12, 1982).

11. See "How Stable Is Our Vacuum?" by Piet Hut and Martin Rees, *Nature* 302 (April 7, 1983).

12. In *Our Final Century: Will Civilisation Survive the Twenty-First Century?* (Arrow Books, 2003), Martin Rees writes: "Hut and I concluded that empty space cannot be so fragile that it can be ripped apart by anything that physicists could do in their accelerator experiments. If it were, then the universe would not have lasted long enough for us all to be here. However, if these accelerators became a hundred times more powerful—something that financial constraints still preclude, but which may be affordable if clever new designs are developed—then these concerns would revive, unless in the meantime our understanding has advanced enough to allow us to make firmer and more reassuring predictions from theory alone." One such clever new design is called a "plasma wake field," which accelerates particles to extraordinary energies over very short distances. If perfected, accelerators could be built at a fraction of the size of today's behemoths.

13. See, for example, "Catching Rays with Radiation Man," by Justin Berton, *East Bay Express*, August 27, 2003. Wagner was also featured in a 1977 *People Magazine* article describing a courtship gone bad: "Walter Wagner's Bizarre Courtship," by Al Donner and Dan Walters, *People*, September 26, 1977. Wagner strongly disputes *People*'s version of those events. See also "Uranium Hunter Follows Trail of Tiles," by Alan Boyle, May 30, 2003, www.msnbc.msn.com/id/3077213/.

14. *Becoming a Critical Thinker: A Guide for the New Millennium*, 2nd ed., by Robert Todd Carroll, Pearson, 2005.

15. See "Problems with Empirical Bounds for Strangelet Production at RHIC," by A. Kent, available at arXiv, dated September 10, 2000, article ID

hep-ph/0009130, and "A Critical Look at Risk Assessments for Global Catastrophes," by A. Kent, *Risk Analysis* 24 (2004): 157–168, available at arXiv, dated December 10, 2003, article ID hep-ph/0009204.

16. See "Might a Laboratory Experiment Destroy the World?" by F. Calogero, *Interdisciplinary Science Reviews* 25, no. 3 (2000): 191.

Chapter 9

1. The Higgs will be difficult to spot at the LHC if its mass is less than around 140 GeV, because a Higgs particle with a mass in that region is most likely to decay into a bottom quark and an anti-bottom quark, and these particles are produced in vast numbers in almost every collision in the machine. These quarks effectively swamp any excess of quarks created from a decaying Higgs particle.

2. Six major improvements were made to LEP to boost its performance in 1999 and 2000. For a detailed explanation of each and its effect on beam energy, see "Direct Search for the Standard Model Higgs Boson," by P. Janot and M. Kado, *Comptes Rendus Physique* 3 (2002): 1193.

3. For a more extensive account, see the CERN document "The Aleph Experience: 25 Years of Memories," 2nd ed., January 2006.

4. Most of the damage was to plastic insulation around cabling. See "Absorbed Doses and Radiation Damage During the 11 Years of LEP Operation," by H. Schönbacher and M. Tavlet, *Nuclear Instruments and Methods in Physics Research B* 217, no. 1 (2004): 77–96.

5. This first Higgs-like signature was recorded by the Aleph detector on June 6, 2000, as part of LEP run 54698.

6. See minutes of the 55th meeting of the LEP Experiments Committee, July 20, 2000.

7. See minutes of the LEP Committee Special Seminar, closed session, September 5, 2000.

8. See "CERN Considers Chasing Up Hints of Higgs Boson," by Alison Abbott, *Nature* 407 (September 7, 2000).

9. See LEP Committee, Special Seminar, minutes.

10. See minutes of the 148th meeting of the CERN Research Board, September 14, 2000.

11. The minutes of the 148th meeting of the CERN Research Board, September 14, 2000, state that "running LEP in 2001 would delay the LHC by 18 months with an estimated cost of about 100m[illion] Swiss francs."

12. See S. W. Hawking, "Virtual Black Holes," *Physical Review D* 53, no. 6 (1996): 3099–3107, available at arXiv, dated October 16, 1995, article ID hep-th/9510029v. In the article's abstract, Hawking states: "This loss of quantum

coherence is very small at low energies for everything except scalar fields, leading to the prediction that we may never observe the Higgs particle."

13. See "Clash of the Atom-Smashing Academics," by Alastair Dalton, *The Scotsman*, September 2, 2002.

14. For a detailed discussion of the different decays and probabilities, see "Direct Search for the Standard Model Higgs Boson," by P. Janot and M. Kado, *Comptes Rendus Physique* 3 (2002): 1193.

15. Ross Berbeco, speaking on *Frontiers*, BBC Radio 4, November 1, 2000.

16. See minutes of the 56th meeting of the LEP Experiments Committee, November 3, 2000.

17. Patrick Janot, quoted in "On a Particle's Trail, Physicists Seek Time," by James Glanz, *New York Times*, November 4, 2000.

18. The minutes of the 149th meeting of the CERN Research Board, November 7, 2000, say that "after a long and arduous discussion, which took into account the memo from the spokesmen of the LEP collaborations, the members of the research board could not agree on a recommendation to the Director General."

19. See notes of the restricted meeting of the European Committee for Future Accelerators, held at DESY, Germany, November 30, 2000.

20. CERN, press release, November 8, 2000.

21. See "CERN Split over Collider Closure," *Physics World*, December 1, 2000.

22. Email correspondence from S. W. Hawking.

23. See "CERN's Head Rejects Mismanagement Claims," by Alison Abbott, *Nature* 413 (October 18, 2001).

24. See "Search for the Standard Model Higgs Boson at LEP," *Physics Letters B* 565 (2003): 61.

25. The article, "No Sign of the Higgs Boson," published on December 5, 2001, began: "The legendary particle that physicists thought explained why matter has mass probably does not exist."

Chapter 10

1. An idea suggested by Leon Lederman, former director of Fermilab.

2. Conway wrote four detailed accounts of this episode on the Cosmic Variance blog, hosted by *Discover* magazine. Two appeared on January 26, 2007, with follow-up articles on March 9 and October 22.

3. For an introduction to supersymmetry, see, for example, "Nature's Blueprint: Supersymmetry and the Search for a Unified Theory of Matter and Force," by the Fermilab scientist Dan Hooper, *Smithsonian*, September 2008. For a more in-depth guide, see *The Quantum Theory of Fields: Supersymmetry*, by Steven Weinberg, Cambridge University Press, 2000.

4. Dorigo writes an insightful and entertaining blog called "A Quantum Diaries Survivor" at Scientificblogging.com. The article referred to appeared on January 19, 2007.

5. See "Glimpses of the God Particle?" *New Scientist*, March 3, 2007.

6. See Sean Carroll's blog, "What Will the LHC Find?" at Cosmic Variance, August 4, 2008.

7. See "If LHC Is a Mini-Time-Machines Factory, Can We Notice?" by Andrey Mironov et al., *Facta Universitatis. Series: Physics, Chemistry and Technology* 4 (2006): 381–404, available at arXiv, dated October 17, 2007, article ID hep-th 0710.3395v1, and "Time Machine at the LHC," by I. Aref'eva and I. Volovich, *International Journal of Geometric Methods in Modern Physics* 5, no. 4 (2008): 641–651, available at arXiv, dated October 26, 2007, article ID hep-th 0710.2696v2, and "2008: Does Time Travel Start Here?" by Michael Brooks, *New Scientist*, February 9, 2008.

8. CERN compiled an extremely detailed account of the incident called "Report of the Task Force on the Incident of 19 September 2008 at the LHC," by M. Bajko et al., LHC Project Report 1168, March 31, 2009.

9. A lot of fuss was made of Nielsen and Ninomiya's theory when it was brought to the attention of the wider public by an equally mischievous essay by Dennis Overbye in the *New York Times*: "The Collider, the Particle and a Theory About Fate," October 12, 2009. In one of his papers, "Imaginary Part of Action, Future Functioning as Hidden Variables" (available at arXiv, dated November 20, 2009, article ID quant-ph 0911.4005v1), contributed to the Quantum Theory: Reconsideration of Foundations conference held at Vaxjo University, Sweden, June 14–18, 2009, Nielsen states: "We claim in our model that the SSC (Superconducting Supercollider) was stopped by the US Congress due to backward causation from the big amounts of Higgs particles, which it would have produced, if it had been allowed to run." Many physicists could not work out whether Nielsen and Ninomiya were serious. In an email exchange with Nielsen, he told me: "It is really not a theory and we do not really believe it ourselves."

Chapter 11

1. At the time of my visit, Fermilab had recently published three articles on the hunt for the Higgs particle, one from each of the detector teams, CDF and DZero, and a third that combined the results from both. The paper that shows the data for a potential 115-GeV-mass Higgs particle is from the CDF group T. Aaltonen et al., "Inclusive Search for Standard Model Higgs Boson Production in the WW Decay Channel Using the CDF II Detector," *Physical Review Letters* 104 (2010), available at arXiv, article ID 1001.4468v2. For more on

signs of a 115-GeV Higgs at Tevatron, see "New Tevatron Higgs Limits Got Worse, but the 115 GeV Excess Is Growing!" by Tomasso Dorigo at "A Quantum Diaries Survivor" blog, November 19, 2009.

2. See "Endgame for the Tevatron," by John Conway, Cosmic Variance blog, September 21, 2009.

3. See CERN symposium, "From the Proton Synchroton to the Large Hadron Collider—50 Years of Nobel Memories in High-Energy Physics," December 3–4, 2009. Veltman gave his lecture, "The LHC and the Higgs Boson," on the second day. Video and slides available from the CERN website.

4. See "The Unhiggs," by David Stancato and John Terning, *Journal of High Energy Physics*, no. 11 (2009): 101.

5. The phrase and variations on it are commonly attributed to both Niels Bohr and Yogi Berra. Bohr's version appears in, among other works, *Between Inner Space and Outer Space: Essays on Science, Art, and Philosophy*, by John Barrow, Oxford University Press, 2000.

6. For more on the idea of manipulating the Higgs field, see an online seminar by physicist Kim Griest at the University of California, Davis. The seminar, part of a series called "Atoms to X-Rays: The Mystery of Empty Space: Higgs Bosons, Vacuum Energy and Extra Dimensions," can be found on the UCSD (University of California at San Diego) TV website at www.ucsd.tv.

7. In *A Zeptospace Odyssey: A Journey into the Physics of the LHC*, by Gian Francesco Giudice, Oxford University Press, 2010, the author states: "It could be done only by heating the universe to temperatures above 10^{15} degrees, a value one hundred million times larger than the temperature in the centre of the sun. It is very unlikely that these enormous temperatures will be ever attained during the existence of humanity, even under the most pessimistic extrapolations of global warming."

8. Several authors have written papers on the hidden-worlds idea. For a good introduction, see Chapter 12, "Higgs Bosons of a Hidden World," and references therein, in James Wells's *Lectures on Higgs Boson: Physics in the Standard Model and Beyond*. The paper is available online at arXiv, dated September 25, 2009, article ID hep-ph/0909.4541v1.

BIBLIOGRAPHY

Aleph collaboration. "The Aleph Experience." CERN report, January 2006.

Ananthaswamy, A. "Glimpses of the God Particle." *New Scientist*, March 2007.

Araf'eva, I. Ya., and I. V. Volovich. "Time Machine at the LHC." *International Journal of Geometric Methods in Modern Physics* 5, no. 4 (2008): 641–651. Available at arXiv, article ID 0710.2696.

Arnaudon, L., et al. "Effects of Terrestrial Tides on the LEP Beam Energy." CERN report, March 1994.

Bajko, M., et al. "Report of the Task Force on the Incident of 19 September 2008 at the LHC." CERN Project Report 1168, 2008.

Berton, J. "Catching Rays with Radiation Man." *East Bay Express*, August 2003.

Brooks, M. "Does Time Travel Start Here?" *New Scientist*, February 2008.

Calogero, F. "Might a Laboratory Experiment Destroy Planet Earth?" *Interdisciplinary Science Reviews* 25, no. 3 (2000): 191–202.

Cashmore, R., and C. Sutton. "The Origin of Mass." *New Scientist*, April 1992.

Cline, D. *Weak Neutral Currents: The Discovery of the Electro-Weak Force.* Westview Press, 1997.

Close, F. *Antimatter.* Oxford University Press, 2009.

Crease, R. "Case of the Deadly Strangelets." *Physics World*, July 2000, 19–20.

Cropper, William H. *Great Physicists: The Life and Times of Leading Physicists from Galileo to Hawking.* Oxford University Press, 2001.

Ellis, J., et al. "A Phenomenological Profile of the Higgs Boson." *Nuclear Physics B* 106, no. 2 (1976): 292–340.

Ellis, J., et al. "Review of the Safety of LHC Collisions." *Journal of Physics G* 35, no. 11 (2008): 115004–115021.

Englert, F., and R. Brout. "Broken Symmetry and the Mass of Gauge Vector Bosons." *Physical Review Letters* 13, no. 9 (1964): 321–323.

Fancey, N. "Interview with Peter Higgs." *Physics Education* 33, no. 1 (1998): 63–65.

Farmelo, Graham. *The Strangest Man: The Hidden Life of Paul Dirac, Quantum Genius.* Faber and Faber, 2009.

Giudice, Gian Francesco. *A Zeptospace Odyssey: A Journey into the Physics of the LHC.* Oxford University Press, 2010.

Glashow, S. "Towards a Unified Theory: Threads in a Tapestry." Nobel Lecture, December 1979.

Glashow, S., and R. Wilson. "Nuclear Physics: Taking Serious Risks Seriously." *Nature,* December 1999.

Gould, S. J. "Ice-Nine, Russian Style." *New York Times,* August 30, 1981.

Greene, B. *The Elegant Universe: Superstrings, Hidden Dimensions, and the Quest for the Ultimate Theory.* Vintage Books, 1999.

Guralnik, G., C. R. Hagen, and T. Kibble. "Global Conservation Laws and Massless Particles." *Physical Review Letters* 13, no. 20 (1964): 585–587.

Halpern, P. *Collider: The Search for the World's Smallest Particles.* Wiley, 2009.

Hawking, S. W. "Virtual Black Holes." *Physical Review D* 53, no. 6 (1996): 3099–3107. Available at arXiv, article ID hep-th/9510029.

Higgs, P. W. "Broken Symmetries, Massless Particles and Gauge Fields." *Physics Letters* 12, no. 2 (1964): 132–133.

———. "Broken Symmetries and the Masses of Gauge Bosons." *Physical Review Letters* 13, no. 16 (1964): 508–509.

———. "Spontaneous Symmetry Breakdown Without Massless Bosons." *Physical Review* 145, no. 4 (1966): 1156–1163.

Hoddeson, Lillian, Laurie Brown, Michael Rioran, and Max Dresden, eds. *The Rise of the Standard Model: Particle Physics in the 1960s and 1970s.* Cambridge University Press, 1997.

Hoddeson, Lillian, Adrienne W. Kolb, and Catherine Westfall. *Fermilab: Physics, the Frontier and Megascience.* University of Chicago Press, 2008.

Jaffe, R. L. "Review of Speculative 'Disaster Scenarios' at RHIC." Brookhaven National Laboratory report, July 2000.

Jammer, M. *Concepts of Mass in Contemporary Physics and Philosophy.* Princeton University Press, 2000.

Kane, G. "The Dawn of Physics Beyond the Standard Model." *Scientific American,* June 2003.

———. "The Mysteries of Mass." *Scientific American,* July 2005.

Konopinski, E., et al. "Ignition of the Atmosphere with Nuclear Bombs." Declassified technical report, Los Alamos National Laboratory, 1946.

Krige, J. "Distrust and Discovery: The Case of the Heavy Bosons at CERN." *Isis* 92, no. 3 (2001): 517–540.

Leake, J. "Big Bang Machine Could Destroy Earth." *Sunday Times,* July 18, 1999.

Lederman, L., and D. Teresi. *The God Particle: If the Universe Is the Answer, What Is the Question?* Mariner Books, 2006.

Llewellyn-Smith, C. "How the LHC Came to Be." *Nature,* July 2007.

Maddox, M. "The Case for the Higgs Boson." *Nature,* April 1993.

Mironov, A., et al. "If LHC Is a Mini-Time-Machines Factory, Can We Notice?" *Facta Universitatis. Series: Physics, Chemistry and Technology* 4 (2006): 381–404. Available at arXiv, article ID 0710.3395v1.

Pais, A. *Inward Bound: Of Matter and Forces in the Physical World.* Oxford University Press, 1986.

Quigg, C. "Spontaneous Symmetry Breaking as a Basis of Particle Mass." *Reports on Progress in Physics* 70, no. 7 (2007): 1019–1054.

———. "The Coming Revolutions in Particle Physics." *Scientific American,* February 2008.

Randall, L. *Warped Passages: Unravelling the Mysteries of the Universe's Hidden Dimensions.* Allen Lane, 2005.

Rees, M. *Our Final Century: Will Civilisation Survive the Twenty-First Century?* Arrow Books, 2003.

Salam, A. "Gauge Unification of Fundamental Forces." Nobel Lecture, December 1979.

———. *Unification of Fundamental Forces.* Cambridge University Press, 1990.

———. "The Role of Chirality in the Origin of Life." *Journal of Molecular Evolution* 33, no. 2 (1991): 105–113.

Schopper, H. *LEP: The Lord of the Collider Rings at Cern, 1980–2000.* Springer, 2009.

Shears, T. G., et al. "In Search of the Origin of Mass." *Philosophical Transactions of the Royal Society,* October 2006.

Smolin, L. *The Trouble with Physics: The Rise of String Theory, the Fall of Science and What Comes Next.* Allen Lane, 2006.

Steinhardt, P. J., and N. Turok. *Endless Universe: Beyond the Big Bang.* Weidenfeld and Nicolson, 2007.

Stewart, I. *Why Beauty Is Truth: A History of Symmetry.* Basic Books, 2007.

't Hooft, G. "A Confrontation with Infinity." Nobel Lecture, December 1999.

Veltman, M. "From Weak Interactions to Gravitation." Nobel Lecture, December 1999.

Weinberg, S. "A Model of Leptons." *Physical Review Letters* 19, no. 21 (1967): 1264–1266.

———. *The First Three Minutes: A Modern View of the Origin of the Universe.* Basic Books, 1977.

———. "Conceptual Foundations of the Unified Theory of Weak and Electromagnetic Interactions." Nobel Lecture, December 1979.

———. *Dreams of a Final Theory: The Scientist's Search for the Ultimate Laws of Nature.* Vintage Books, 1992.

———. *Facing Up: Science and Its Cultural Adversaries.* Harvard University Press, 2001.

———. "From BCS to the LHC." *Cern Courier*, January 2008.

Wilczek, F. "Masses and Molasses." *New Scientist*, April 1999.

———. "Mass Without Mass: Part I." *Physics Today*, November 1999.

———. "Mass Without Mass: Part II." *Physics Today*, January 2000.

———. "The Origin of Mass." MIT annual paper, 2003.

———. "Asymptotic Freedom: From Paradox to Paradigm." Nobel Lecture, December 2004.

———. "In Search of Symmetry Lost." *Nature*, January 2005.

———. *The Lightness of Being.* Basic Books, 2008.

Woit, P. *Not Even Wrong: The Failure of String Theory and the Continuing Challenge to Unify the Laws of Physics.* Jonathan Cape, 2006.

INDEX